natural beauty
Homemade Recipes for Radiant Skin & Hair

natural beauty

Homemade Recipes for Radiant Skin & Hair

elizabeth tenhouten

hatherleigh

⟫ hatherleigh

Hatherleigh Press is committed to preserving and protecting the natural resources of the earth. Environmentally responsible and sustainable practices are embraced within the company's mission statement.

Visit us at www.hatherleighpress.com and register online for free offers, discounts, special events, and more.

Natural Beauty
Text copyright © 2013 Elizabeth TenHouten

Library of Congress Cataloging-in-Publication Data is available upon request.
ISBN: 978-1-57826-446-9

Cover and Interior Design by DcDesign
Set in Berkeley

Printed in the United States

10 9 8 7 6 5 4 3 2 1

"Since love grows within you, so beauty grows.
For love is the beauty of the soul."

—St. Augustine

"Elizabeth's beauty recipes are the foundation upon which beautiful skin is built. Follow with a primer for perfect skin!"
—DAVIS FACTOR, Co-Founder of Smashbox Cosmetics
and Smashbox Studios

"I adore Elizabeth TenHouten. Elizabeth's beauty is not just skin deep, though it certainly doesn't hurt that her recipes make sure her skin is beautiful, too. As a gal on the go, I appreciate how easy and fun her treatments are. I rarely have to look further than my own kitchen cabinet, and that's refreshing."
—ELAINE HENDRIX, actress and animal advocate

"Just be your naturally beautiful self with the pure and simple beauty recipes in this book!"
—EDEN SASSOON, Owner of EDEN by Eden Sassoon

"From the jungles of Vietnam to the jungles of Beverly Hills, I know the value of natural beauty remedies. I love Elizabeth, and love her approach to natural beauty!"
—KIM VO, Master Colorist to A-list Hollywood (including Britney Spears,
Goldie Hawn, Kate Hudson, and Pamela Anderson), judge on the hit
show "Shear Genius" on Bravo, also seen on E! News Entertainment

"I think some of the best health and skin care remedies are the ones that we have learned from our grandmothers. Elizabeth TenHouten brings those old-fashioned remedies back from the Dark Ages and makes them in vogue again…and the best part is they are affordable and they work!!"
—LORI LOUGHLIN, actress

"TenHouten has tapped into a growing beauty trend DIY. *Natural Beauty* is the perfect beginning in the art of creating at-home beauty and spa treatments for all skin types. It's easy to read, fun to follow, and inspires the inner cosmetic chemist in all of us."

—LYNN LUDLAM, President of Beauty Industry West

"I love Elizabeth's natural beauty recipes because you have to start with a great foundation. If you don't have a great canvas, you can't create a beautiful face. Less is always more so that your natural beauty shines through, and Elizabeth's recipes allow your skin to radiate naturally."

—THEA BELLA ISTENES, Celebrity Makeup Artist for Smashbox Cosmetics, Allure, and *Elle* magazine, clients include Eva Longoria, Kate Bosworth, Melissa Miller, and Miranda Kerr

"Whether you're in Valentino or the Gap, the first part of putting together any look is making sure your skin is glowing and your hair is soft and shiny. Elizabeth's beauty recipes are an easy way for any woman to give herself that extra va-va-voom factor! I don't let a week go by without using one of her beauty recipes. I will be giving my clients copies of this book the minute it hits the shelves!"

—KAMALA BERNSTEIN, Stylist for L'Uomo Vogue, clients include Mary J. Blige, Robert DeNiro, and Nicolette Sheridan

"Elizabeth TenHouten is a beauty/skin care expert, and a huge believer in nature's ability to keep you young—both inside and out."

—LAUREL HOUSE, Editor for Discovery Channel's Planet Green

"Beauty expert Elizabeth TenHouten is here for the do-it-yourself at-home beauty enthusiast. Her book is filled with at-home tips/tricks and recipes for all-natural beauty remedies to achieve A-lister skin naturally!"

—WES FERGUSON, YoungHollywood.com

"The beauty recipes in Elizabeth's book are the perfect path to a glowing complexion by using all-natural ingredients. I also love her philosophy of beauty radiating from within. I highly recommend this book to anyone looking to for a fun, natural beauty regimen."

—JENNIFER NICHOLSON, fashion designer and owner of Pearl Drop

"A simple, effective, and natural approach to using foods for maximizing skin health and beauty. Elizabeth is very knowledgeable in her field and offers readers a wealth of her knowledge base. This book is a must read!"

—GUY LANGER, Society of Cosmetic Chemists (SCC) California Education Chair, two-time winner of the SCC Chapter Merit Award

"Elizabeth TenHouten truly knows about beauty from the inside out! Her passion, knowledge, creativity, and know-how shine through in every fabulous recipe! Elizabeth has been a guest on my radio and television shows, and every time I learn another great recipe using natural foods right in my kitchen, like her luxuriating brightening mask! Pick up this book and get beautiful from the inside out!"

—LISA DAVIS, MPH, C.N.C. Creator, Host, and Producer of *It's Your Health Radio* and *It's Your Health TV* and Producer and co-host of the radio show, *Beauty Inside Out with David Pollock*

"As a yogi, I value the natural approach to beauty that Elizabeth offers in this inspirational book. Teaching yoga makes my soul glow and Elizabeth's beauty recipes make my skin glow!"

—BRIGITTE BEDI, Yogi at Maha Yoga Studio

Acknowledgments

I WOULD LIKE TO recognize Andrew Flach, Publisher of Hatherleigh Press, for his continued belief in my work and purpose. I extend a warm acknowledgment to Anna Krusinski and Ryan Kennedy for their thoughtful and insightful edits. A special thank you to Ryan Tumambing for all of his hard work and DCDesigns for their beautiful design. I am grateful for the support of all the people at Hatherleigh Press and Random House.

I very much appreciate my friends and family for their unfailing encouragement; and especially thank Douglas Cohen for his support of my dreams. I am also grateful to my mother for believing in me and inspiring me to grow as a person.

Contents

Introduction

A WORLD OF BEAUTY is within your grasp, waiting to be discovered in your very own pantry. All of Mother Nature's pure, all-natural ingredients are waiting to become a part of your beauty routine.

The natural beauty recipes in this book are intended to effortlessly resolve all of your beauty woes, from under-eye darkness to chapped lips. But the real beauty of this book is that it addresses a wide range of topics, from the simplest of undesirable skin conditions (such as dry skin) to more complex skin issues (like rosacea) all through a healthful, wholesome, all-natural approach. No harsh chemicals; no harmful, unidentifiable ingredients—nothing but natural remedies to nourish your skin and leave it healthy and beautiful.

Knowing exactly what ingredients are involved in the creation of each beauty recipe gives you the peace of mind that your skin, hair, and body are in nature's hands. And certainly, it's important that each recipe has a shelf life. Whether it's a protein hair mask for damaged strands or a hydrating "feet treat" for the lady who loves her high heels, there are absolutely no preservatives in any of the beauty recipes in this book. Additionally, the step-by-step instructions explain why each specific ingredient is beneficial in its own way, as well as how it reacts with the other ingredients to provide you with the full beautifying benefits. Even more valuable still is the simple, intuitive nature of the recipes themselves so that you can easily follow in my footsteps and make your *own* natural, homemade beauty recipes for radiant skin and hair.

Our skin is the largest organ in our body, but it has many delicacies and complexities to which the beauty recipes in this book cater and respect. For instance, the skin under your eyes is sensitive because near your eyes you have tiny mucus membranes, which are actually quite vulnerable. The skin under your eyes can't even be moisturized with the same mixture that hydrates the stronger skin on the rest of your face. And clearly, the delicate skin under your eyes and the skin on your face are different from the skin on the rest of your body. For example, your legs, which likely undergo some form of hair removal (such as shaving or waxing) on a regular basis, require a more aggressive beauty recipe to replenish their moisture;

in fact, it's ideal to utilize beauty recipes that will actually help your skin retain its own natural oils and moisture. All of this is possible with the simple DIY natural beauty recipes in this fun book! Let *Natural Beauty* be your glamorous go-to guide for all-natural skin and hair care.

Although this book absolutely addresses the changes our skin undergoes during the changing of the seasons, it also wisely acknowledges that cementing a label or permanent typology of skin (for example, oily skin or dry skin) fails to appreciate the environmental factors that account for our skin's constantly evolving needs. The focus of this book is to teach you how to assess and reassess your skin's condition on an ongoing basis. This ensures that you will always stay informed of your best route to a glowing complexion. Really get to know your skin, and stay in touch with it. The seasons change, and so do your skin and your hair.

A beauty redefined by nature's healing forces and the virtues of pure ingredients is the path to wellness. With my love of food, and as a natural cook, I yearned to fuse these loves of mine into something greater. It has blossomed, beautifully, into a greater fundamental truth, and it now encompasses my understanding of beauty.

In addition to 75 all-natural homemade recipes for radiant skin and hair, this helpful beauty book offers a chapter devoted to describing skin issues and providing simple DIY solutions. So, indulge in your own luxury at-home spa and reach for your own natural beauty!

natural beauty

1

A History of Reaching
for Beauty

WOMEN HAVE INDULGED in forms of beauty modification for thousands of years, walking the same paths even as their methods improved. These beauty modifications range from (but are not limited to) intricately designed tattoos, scarification for beauty, teeth filing, and intentional weight gain (in cultures where the most beautiful woman is the fattest) or conversely, starving oneself to achieve the slimness and low weight idealized in Western cultures. Each of these ancient (and not-so-ancient) seemingly primitive ways of reaching for beauty has its parallel in modern times. It continues even today, with improvements and technological advancements geared toward obtaining the same results desired in centuries past.

In fact, many ancient cultures would gasp at our modern advances in methods of attaining beauty. They might even find these new methods barbaric in comparison to their simpler approaches.

I hope to bridge the gap and find a commonality among all cultures in their quests for beauty. It's most important to understand how we, as women, as beings, are alike, rather than focusing on our differences. If we are to suspend the sorts of judgments that limit us, then it's pivotal to focus on the similarities that bind us. The very act of reaching for beauty itself is natural, and the nature of the pursuit and methods used should not minimize your value as a being of natural beauty. Reaching enlightenment on the issue of natural beauty is about suspending judgment and recognizing the beauty within all beings. I think that a woman shouldn't feel plastic if she has undergone plastic surgery. Let's all suspend judgment, please, including self-judgment.

It's a waste of time to stand on a metaphorical high horse and judge others, to claim that with modern advancements women are seeking a fake, or artificial, standard of beauty. So long as no one is harmed by her methods, then there is little difference between the modern beauty seeker and her long line of predecessors on the path. The only harm possible in this scenario comes from the judgment

natural beauty

of others. The one who is harmed is the one who wastes time judging others. It's always best to love someone as they are, even if that means overcoming a personal difficulty in suspending judgment. It's an opportunity for growth in your own soul. People will become who they are organically. It's the natural way of the evolving self. It's the existential concept that we are always in a state of becoming, always realizing new aspects of our potential selves.

Beauty enhances the quality of life, for humans and animals alike. Animals instinctively reach for beauty in their own naturalistic way. Bowerbirds, for example, reach for beauty in their nests, which they take weeks to create. These incredible birds create an enticing nest filled with colorful berries and unique and aesthetically pleasing leaves, nuts, flowers, and fungi to represent themselves as a beautiful and suitable mate. It's quite sweet, actually; the care that these industrious winged lovers put into their romantic encounter is heartwarming. There are countless, fascinating ways that animals reach for beauty. This example from the animal kingdom indicates that the desire for beauty extends beyond humans and is truly an inherent aspect of life itself. The bowerbirds find beauty through the inclusion of natural elements in the same way that the natural ingredients in these beauty recipes will for you.

Throughout history, this deep-seated need has played a huge role in humanity's development and the shape of our societies. Everything from cultural expression and religious ideals (as with the ancient Egyptians) to social class and status (as with the bourgeoisie of pre-Revolutionary France) has revolved around becoming the best person you can be, around becoming beautiful.

To reach for beauty is to reach for personal growth; to enhance one's beauty is to enhance one's sense of self. And because humans, more so than any other creature on Earth, strive to improve themselves and their world, it's no surprise that our tradition of improving ourselves, of increasing our beauty is ancient, storied, varied, and constant.

Historical and Modern Parallels in Reaching for Beauty

Hair	Marie Antoinette adorned her hair with dramatic ornaments such as model ships, flowers, and feathers. She created a statement with powdered wigs more than two feet high reinforced with wires. Victorian women wore exposed hats to allow their hair to be lightened by the sun. It would lighten to a strawberry shade, known as Victorian blonde. Babylonian men sprinkled gold dust in their hair to appear more desirable.	Pop icon Lady Gaga creates elaborate "hair costumes" with lace and pearls for artistic appeal. Popular hair-coloring products are showcased on television commercials, recruiting famous models to sell their product for the perfect hair color. It's all the rage among tweens to attach strands of feathers and glittered hair extension strands in their tresses.
Teeth	False teeth were invented using ivory and white marble, widely worn all over Europe. George Washington was rumored to have worn wooden teeth as dentures.	Teeth whitening products are sold at your local market or pharmacy along with whitening toothpaste. Veneers and teeth capping are common for achieving a beautiful, aesthetic smile.
Skincare	Chinese women of the first century ingested a special powder to lighten their complexions.	Intense pulsed light lasers and photo facials promise to remove sunspots or dark freckles.
Face Makeup	Queen Elizabeth I famously painted her face with white lead and vinegar, creating a homemade foundation.	Max Factor invented the first natural-looking foundation in 1914 originally called "flexible greasepaint" for the movie industry. Starlets raved about it.
Body Hair Removal	In 200 BCE, during the wild, sexual rule of Caligula in Rome, women used seaweed to remove nipple hair.	Estheticians have created entire careers from waxing and threading as a form of hair removal. Laser hair removal is commonplace.
Eye Makeup	Cleopatra made her own eyeliner from kohl and mesdemet.	Young girls start experimenting with eye makeup at an early age, and eye makeup is a mark of womanhood.
Jewelry	In some parts of Africa, women wear gold rings to elongate their beautiful necks.	Necklaces enhance outfits. Most women have their ears pierced and even other body parts.
Aroma	Egyptians in the fifteenth-century BCE made perfume out of plant extracts to enhance their beauty presence.	It has become popular to create your own customized perfume according to your pH balance, a service available at many department stores.
Eyebrows	Israeli women used threading to keep their thick eyebrows shaped beautifully.	Eyebrow guru (yes, in our modern times there is a woman famous for doing eyebrows!), Anastasia, has created special shaping guides for women to pluck their eyebrows to perfection.

2

<figure>✳</figure>

Skin Typology

O UR SKIN CAN generally be classified as one of three skin types: normal, oily, or dry. Beyond this rudimentary understanding of skin types, remember that your skin is an evolving organ, and you should assess it regularly. Having said that, you can understand skin on a basic level by understanding these three categories of skin types.

Normal, oily, and dry skin are the three most generally recognized skin types and can also encompass skin conditions such as sensitive skin, which are not to be confused with the generality of sensitive skin as a skin typology. In chapter 8, I address different skin conditions or "woes," but these occur outside of the basic skin type descriptions. Please respect that there is no such thing as a consistent skin type and that, just like the seasons, your skin is ever evolving and shedding as it reacts to environmental and internal stimuli.

Therefore, when reading these classifications of skin, please note that you may experience all three of them several times throughout your life. I want to impart the importance and effectiveness of caring for your skin while constantly reevaluating its present type. In other words, I really don't want you to classify your skin in a permanent sense. The healthiest way to beautiful skin is to monitor and assess what your skin type is in the moment and care for it accordingly.

If your skin presents as normal, this means that you have an even, monochromatic skin tone without melasma (evident darker pigmented spots). Normal skin has a healthy hue and minimal pores and responds well to gentle beauty recipes. For instance, care for normal skin includes gentle exfoliation such as my Gentle Exfoliant with sour cream and almond and coconut oils (see page 104).

Oily skin presents as shiny, most often in the T-zone, which includes the forehead and the area between the eyebrows, nose, and chin. Marked by larger pore size, oily skin has the tendency to acquire blackheads or whiteheads as the pores become clogged with dirt and oil. Care for oily skin is slightly more involved in terms of cleansing. It's wise to also follow cleansing with a toner and a lightweight moisturizer to restore pH balance. A great natural beauty recipe for oily skin is my Fruit Acid Exfoliant (page 105), which includes naturally antibacterial ingredients such as euca- lyptus honey to keep your skin clear. Another great recipe for oily skin is my Clarifying Clay recipe (page 97), an intensely clarifying exfoliant. Oily skin, however, is less inclined to wrinkle as compared to normal skin or the next typology of skin, dry skin.

Dry skin presents as rougher to the touch and often appears dull or lackluster. If you are experiencing dry skin, my Brightening Mask (page 108) is in order. The natural lime juice provides the brightening action of vitamin C for a radiant glow. Dry skin craves moisture, so be sure to generously moisturize and use sunscreen, as always. Sunscreen is particularly valuable if your skin is drier because the damaging effects of the sun show in the form of fine lines and wrinkles appearing more readily.

Again, I don't subscribe to the notion of a typology of skin. Although your skin may at times be oily or dry, you are not sen- tenced to the phrases *oily skin* or *dry skin* or any other terminology that embodies a permanent state of imperfection. So, before you sew that scarlet letter *O* for "oily" on all your washcloths, let's understand the *continuous evolution* of our skin.

Our skin is an organ, and it adapts and changes as it copes with internal and external factors. Internal factors include (but are not limited to) hormonal changes, genetic predispositions, and diet. External factors include (but are not limited to) environmental influences, sun damage, toxic stress, weather conditions, sleep deprivation, or bad habits such as smoking or drinking. And so it is that during the winter months your skin may be drier, but that doesn't mean it will be dry forever, nor is your skin simply dry skin. The useless label is permanent, but your skin is not.

Proper skincare requires your checking in and assessing the current state of your skin. Every 28 days a new epidermal layer is revealed, and your skin should be subject to a new evaluation. As skin typology changes with the seasons, environmental changes, and external factors, it's important to change how you care for and beautify your skin. There is no "one size fits all" treatment for your skin; just as your skin changes, you have to adapt and evolve your skincare habits in accordance with the specific changes that are naturally occurring in your life. Once-dry skin from the winter season may now be acne prone, oily skin during summertime. During a woman's menstrual cycle, she may break out with a blemish or pimple; this is a hormonal change and must be understood as such.

Making beauty recipe adjustments is essential. Moisturize more during times of dryness, clarify more during times of humidity, and

natural beauty

brighten more in times of dullness. A simplified natural beauty process that balances your skin is far more beneficial than year-round, chemical-laden products from the beauty counter. Just look in the mirror and "listen to what you see." There is no need to classify your skin when one day it might be drier or more oily than it is the next. The danger of a rigid beauty routine is that you become programmed; you stop "checking in" with your skin. Then, when an occasion arises where your skin needs immediate, special attention, you will ignore it as though you were on autopilot. Remember, your skin loves attention!

Another crucial point to remember is that your everyday, reliable tool for your skin's protection and survival is sunscreen. Think of sunscreen as magic cream! This is the *only* daily form of rigidity that I approve of and, in fact, insist upon. When it comes to using sunscreen, I like to use the phrase "always and forever." Commitment-phobes be at ease; I promise you that sunscreen will never break your heart!

Nature's healing methodology requires little help from expensive chemical products. Nature is on our side. My natural beauty recipes translate to the purity of nature in harmony with efficacious results for radiant skin and hair.

3

Seasons and Beauty

WINTER

Winter is a time of beauty. Snowflakes gracefully fall from the gray skies, the crisp winter wind paints blush on our cheeks—such beauty to behold in this season.

Yet, the winter months can be damaging to your skin for a multitude of weather-related reasons. The wind can be harsh and cause dryness as well as chapped lips and, in certain parts of the country, even chafe the skin on your face, particularly under your nose. So, lip balm is important, like my Sweet Almond Lip Balm (page 93), an excellent moisturizing emollient for your lips. Sweet almond oil helps your skin maintain its own moisture, and this is true of the skin on your lips as well. Try to get a balm with sunscreen in it, such as Blistex, or add your preferred SPF to this beauty recipe.

The sun is hidden by the winter sky but remains capable of giving you a winter sunburn. Remember to wear sunscreen year round. Wintertime calls for hydration because without the heat of the sun making us sweat, it's easy to forget to drink water. A wonderful moisturizing beauty recipe for winter dryness is my Floral Healing Facial (page 113). Aside from having key clarifying ingredients such as carrots rich in vitamin A, it also has honey, oil, and calendula flowers to relieve your skin from inflammation associated with windy climates and to deeply moisturize your skin. Use of heaters in the wintertime can dry your hair out as well. Lustrous Locks (page 64) will hydrate your dry locks in no time. Awareness of these seasonal hazards is the first step toward preparing to have soft, supple skin during the winter months.

SPRING

Springtime has arrived, so remember to stop and smell the tulips. With the spring season upon us, we are still in a period of transitioning from the harsh dryness of winter before our skin is bathed in the warm embrace of the sunlight gifted to us by the new season.

Let's spring forth into the new season with pure and clean skin. Your skin may have called for heavier moisturizers during winter, but it's now time to lighten up and adjust your skincare regimen to suit its new needs. Definitely renew your skin with my Chamomile Steam Facial (page 103) to cleanse your pores and remove any buildup of heavier products you would have used in the past few months. It's also wise to follow up the steam facial beauty recipe with an exfoliation treatment to remove dead skin cells, revealing a brightened complexion. My Papaya Milk Mask (page 106) is ideal because the alpha hydroxy acids in both the milk and the papaya are naturally exfoliating. So, give your skin a little spring-cleaning for a natural glow.

During this seasonal transition, it's important to stay balanced. Maintaining your natural beauty balance relies on your desire to lean inwardly and appreciate your natural self for all that you are worth. An energetic connection exists to show us the way back to beauty into ourselves, naturally. In the midst of the warmth of spring holding us, let's return home to beauty.

"APRIL"

The roofs are shining from the rain,
The sparrows twitter as they fly,
And with a windy April grace,
The little clouds go by.
Yet the backyards are bare and brown,
With only one unchanging tree—
I could not be so sure of spring,
Save that it sings in me.

— SARAH TEASDALE

SUMMER

Summertime is a whimsical time of the year. The sun is shining, inviting you to the beach and to indulge in the warm brightness. Summer is not only a season for less makeup but also a season to opt for multiple applications of SPF protection. The sun is out in all its glory, so you need to be armed with sunscreen to prevent premature aging and sun damage. The most important beauty product you can ever own is sunscreen.

natural beauty

Spending time in swimsuits and even lighter clothing like V-necks instead of turtlenecks can leave your décolletage exposed to sun damage. Although it's important to apply sunscreen to your neck and décolletage in addition to your face, this can be easy to forget. Cucumbers are soothing to irritated, sun-exposed skin. My Cucumber Neck Cream (page 119) provides your neck and décolletage with the relief and rejuvenating properties that this season's skincare calls for.

Your lips can be adversely affected in the summertime, just as they can in the dry winter months. Sunshine feels lovely, but it can blister your lips if overexposed. It's important to apply some sort of lip balm that has an SPF value in it. For instance, I don't leave the house without applying my Blistex with SPF 15. If I wear lipstick, you can rest assured that my SPF Blistex is underneath. It actually conditions your lips, as well, acting like a primer. The summertime is a great time of year to exfoliate your lips. So, indulge in the Sugar Lips natural beauty recipe on page 96 just in time for a summer romance!

Beachy hair is beautiful and natural, but it also needs some summer loving. A basic cleansing rinse will do the trick to return your hair to its more manageable state. My Apple Cider Rinse (page 65) is perfect as a shampoo preparation. Oils from sunscreen may have seeped into your hairline, or your scalp might have some buildup from sweat and conditioner. Both can be remedied with this simple and easy apple cider vinegar rinse.

If your skin feels more oily than usual, you need a clarifying mask to draw out toxins and get your skin back on track. My Clarifying Clay (page 97) beauty mask is perfect for just that. The summer heat can create the appearance of a greasy complexion. This is one reason women switch to a foundation with lighter, more sheer coverage in the summer, and it's why this deeply clarifying mask will leave your skin clearer. Clarifying your skin allows you to continue your fun in the sun.

FALL

During the autumn months, your skin changes with the leaves. You may be recovering from a stubborn sunburn from the previous sunny months; if so, you need to evaluate your delicate skin and perhaps apply some Milky Ice Cubes (page 111) to even out your skin tone and reduce redness and inflammation. If your skin is peeling, don't exfoliate it, but rather apply the milk ice cubes gently to your skin

natural beauty

to help with the exfoliation, as well as with skin tone and inflammation. Whereas the alpha hydroxy acids in milk are clarifying, the beta hydroxy acids are exfoliating,

A creamy moisturizer serves to revitalize your skin. My Honey Almond Moisturizer (page 109) has natural ingredients to actually help your skin retain its own natural moisture while providing extra hydration topically.

4

A Philosophy of Beauty

I S A NATURAL beauty someone who has not enhanced or modified herself in any way, shape, or form since her natural birth? I think that perhaps such a definition is too rigid. Naturally beautiful women have been modifying or enhancing their bodies since the beginning of time. Interestingly, the process of "reaching" for beauty is actually quite an old tradition. This doesn't mean that women have been on operating tables for centuries, but throughout history, they have found methods for attaining what they feel is their own highest form of beauty. Beauty is itself an infinite idea, like a flower that is never done blooming, like a human garden with bulbs blossoming that we didn't know were even in our soil. Let's just continue to water ourselves with love and see what beauty grows!

Now, it may be best to clarify what I mean by "reaching beauty." It's not the *desire* to become your most beautiful self, inside and out. What woman doesn't want to be her most beautiful self? A true natural beauty is a woman who *does* feel beautiful, inside and out. It matters not what others think of any physical enhancements she has made in the name of beauty. We are all reaching, whether you try to erase fine lines by eating antioxidant-rich recipes or applying moisturizers or using injections like Botox or just drinking a lot of water and staying out of the sun.

But how far away must you reach for beauty? It's a challenging query. It provokes an introspective thought process, and I invite you to allow yourself to engage in this way of thinking. As women, we are naturally beautiful and should be proud of our beauty. We should be proud of reaching for beauty and never judge another for her preferred methods.

Our understanding of natural beauty must be more than skin deep. As a beauty and skincare expert, I know that there is a vast array of ways to reach for beauty. Natural beauties reach for beauty via varying methodologies of attaining the desired beautiful result.

For example, for smooth legs, the options range from shaving to waxing to laser-hair removal. It really doesn't matter which avenue you choose to reach your beauty because it's about a personal choice and a personal definition of beauty.

Of course, I don't feel that someone is not a natural beauty simply if she removes the hair on her legs, regardless of the method used. The term *natural beauty* has become co-opted and challenged by the newer, less natural methodologies available to women. With the current popularity of plastic surgery, it's evident that plastic surgery is indeed a viable option to attain the beauty that otherwise might be pursued in a more natural way. If a woman wants to enhance her breasts, there are nonsurgical methods such as topical cream applications, pills that promise a larger bust, pectoral muscle exercises, or simply a padded bra. So, I really don't think that it's fair to judge one woman over another for how she goes about enhancing her beauty. Nor do I believe that such an enhancement dissipates her natural beauty. There is simply no such thing as an unnatural beauty or a phony beauty. Let's celebrate beauty in its entire diverse splendor.

A form encapsulates the act of reaching for beauty, both inwardly and outwardly, that we are naturally inclined to do. The act of reaching for beauty (in whatever form a woman chooses) is met with the abstract concept of finding beauty within the self. The inward return to beauty complements the outward reaching for beauty. As in the infinite model (see Figure A), our reaching and returning collide, and we are met with the reality of a higher truth. And that is natural beauty. Every person is a natural beauty, so let your natural beauty infinitely shine!

And it's important to understand beauty to be an enticing, electric force inside of our spirits. We naturally respond to beauty and to an aesthetically pleasing nature; beauty can recognize itself in others and rejoices in its familiarity. This inner and outer reunion of beauty in one's self and beauty in others is nature's way of taking us "home," of helping us to exist in a state of symbiotic oneness. We are called upon again and again to make a return to beauty. Our existence, our natural state, is one of reaching for beauty.

natural beauty

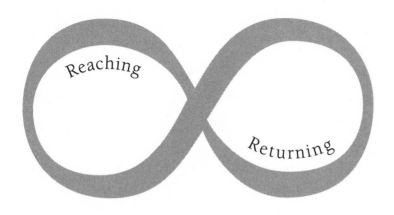

Figure A: The state of returning to, and reaching for, beauty continues infinitely.

This cyclical pattern shown in Figure A continues forever, ad infinitum. What changes in our search are the depths to which we can reach and the heights to which we can return. A deeper sense of self and a heightened level of awareness are the two intricately joined aspects of this truly evolutionary process. Returning to beauty is the act of recognizing the beauty in life. When we see beauty in ourselves and others, when we experience it in nature, we enter into a state of being that is difficult to describe in simple terms.

In the very moment of recognition of beauty, everything becomes enhanced—our selves, our surroundings, our lives as a whole! And the deeper you are in touch with your own reaching for beauty, the more aware you are of the beauty that surrounds you every day, then the more frequently you will find yourself entering into this state of true bliss.

Our fundamental need, beyond the physical needs of the body—shelter, love, sex, and the drive for food—is our basic inclination to connect with beauty. To live a life, to live in a world that is rich, beautiful, and stimulating. It's an essential component to a full life. I believe that reaching for beauty is a basic human drive. I find it to be a poetic wave of back and forth; it's a reaching and finding that never ends. The search goes deeper and holds more meaning the more time and energy you commit. Always ahead of us, just beyond our grasp is the next layer of beauty. We are driven to reach for it, but the core, the pure and true goal, can't be held. It's the beautiful, shining pinnacle of potential beauty actualized.

Becoming is such a beautiful evolution and ought to be dearly embraced as we reach to become more nearly human. This is an existential truth. This theory of reaching for beauty that I raise is a

natural beauty

recurring theme that I wish to lightly introduce as a way of explaining the nature of beauty—that we are, in a sense, born to reach for beauty. It's a natural inclination and is beautiful within itself. So, as we further explore this concept of reaching for beauty, I request that you suspend judgment of others and even of your own process of reaching for beauty. This way, you may comprehend this art of becoming and seeing your natural self. You are a sight of beauty.

This is your glamorous guide to natural skin care, and its results are well within your reach. Your skin has natural needs that can easily be met through a natural approach. Recall that there are many approaches to reaching for one's own beauty. And despite the choices we make, we are all natural beauties by virtue of our inward and outward reaching. The idea that the words *natural* and *beauty* have merged to form a greater meaning, a synthesis of the two, indicates that we now live in a world capable of a far greater, far broader understanding of beauty. What we are given by nature, including our natural inclination to strive for something more beautiful, is the apex by which the two ideas become enmeshed with one another.

My beauty recipes are all natural, at-home recipes for a beautiful, healthy complexion. You *can* do this—so get excited!

5

⁂

Create a Spa in Your Home

THE ATMOSPHERE OF your at-home spa is the most important part of creating a carefree, calming escape from your daily routine. It should really feel like a break in your day-to-day life, a break from the mundane and the obligations you have at work and home. One aspect of creating the ideal ambiance comes from proper lighting. You can't find a level of deep relaxation with blinding florescent lighting. I recommend lighting candles as your source of illumination.

Staying attuned to your body's senses, such as the sense of smell and the sense of light, is another factor in truly allowing yourself to surrender to a beautiful level of calm. An inexpensive and productive way to achieve soothing lighting with a relaxing fragrance is to light a scented candle. These are simple to make yourself, and you can add your favorite fragrance. Of course, you can certainly find scented candles anywhere they are sold, including your local grocery store.

Location is important as well. If you have a bathtub, then I recommend choosing your bathroom as the place to create the meditative realm, where you can reach for beauty in a warm bath filled with bubbles. Perhaps your bedroom window has a lovely view and is better suited to manifesting inner peace. Whatever you choose as your room for your at-home spa, it's important to indulge in the perfect atmosphere that meets your personal needs.

Aside from lighting and aromatic influences, there are yet more things to consider when creating your atmosphere. It's key that you are in a setting where you won't be interrupted. If you are a busy mom, then you should have a conversation with your children that "Mommy needs thirty minutes to herself." You can even have your kids create a pretty sign to post on the door for such occasions. It can be a fun craft activity involving the whole family, using textures like felt letters and glitter, among other childhood favorites. This way, your children won't resent being locked out; rather, they will understand and feel pride in having participated in contributing the sign.

natural beauty

If you have a partner who is there to help make sure you have your moment of meditative, natural beauty reaching by keeping the children occupied, then that is wonderful. But if not, timing is another critical factor in creating the atmosphere to reach for beauty. It may be best to wait until the children are at school or at a friend's house. It's a win-win situation. There is nothing less relaxing for a mom than wondering if the kids are getting along and safely playing. Inhabiting a state of concern is the least desirable atmosphere for anyone, let alone someone reaching for beauty and calm. So, it's important to plan ahead and reserve this important time for yourself.

Not every atmosphere involves the quietude of a closed door, bubble bath, and softly lit candles. Sometimes my at-home spas get a little wild. I like to listen to my favorite playlist on my iPod and rock out while I apply my creams, masks, or exfoliants. Whatever beauty treatment I might be applying, I enjoy a sexy atmosphere with candlelight and Portishead playing in the background. It makes me feel beautiful, and that is the point of an at-home spa. It represents freedom. So, while I also enjoy a luxurious bubble bath with soft lighting and just the sound of my own thoughts if I am meditating, it's nice to know that I have the freedom to reach for beauty in accordance with the direction of my mood. This is all about catering to yourself, and that can't be forgotten.

No matter what kind of atmosphere you aim to create, I believe that the most important part of the atmosphere is cleanliness. Hygiene is *the* first priority when creating your at-home spa. You must be able to wash your skin before applying creams, and you must use clean utensils for mixing and application (as mentioned in the specific Natural Beauty Bytes throughout chapter 7).

Also realize that your at-home spa can also become your own private workout studio. Instead of the gym's preferred radio station, *you* rule the music in your at-home gym. And fear not, for you don't need any actual gym equipment to create your own private gym. If you have an exercise/yoga mat or a clean towel, you are good to go. Perhaps you may want to listen to the news in the background as opposed to inspiring music; either way, your gym must cater to your desires to be effective and leave you wanting more "gym time." On the following pages are several simple mat exercises. These are variations of the sorts of exercises I have engaged in over the course of several years in group and individual exercise classes. Exercise is fantastic for improving circulation, for releasing toxins through sweat, and certainly for clearing your mind. All you need is ten

natural beauty

minutes in the morning, or whenever suits your schedule. If you can't afford a gym membership, you don't have to forego your workout as you reach for beauty through exercise.

AT-HOME GYM

These are my favorite exercises and I hope you like them too! Get your comfy gym clothes on; the only person you need to impress is yourself.

Lovely Leg Lifts

Lay on your side, on your hip. Align your body in a straight line. Flex your bottom foot and anchor it into the ground for resistance and support. Point or extend the foot on top and lift straight up with control and return to a hover, with control. Repeat ten times to conclude one rep. Complete three reps. Repeat this exercise with your other leg.

Lovely Diagonal Leg Lifts

Lay on your side, on your hip. Align your body in a straight line. At the hip, move your legs diagonally at a 45-degree angle. Flex your bottom foot and anchor it into the ground for resistance and support. Point or extend the foot on top, remaining at a 45-degree angle and lift with control. Repeat ten times to conclude one rep. Complete three reps. Repeat this exercise with your other leg.

Dangerous Derrière

Lay on your side, on your hip, in the same position you used for the Lovely Leg Lifts. Bend both knees, keeping your feet near to your bottom. With your feet touching at all times, lift only your knee to the sky and return with control. Repeat ten times to conclude one rep. Complete three reps. Repeat this exercise lying on your other side.

Perfect Plank

Lie on your stomach with elbows bent underneath your shoulders. Come up on the balls of your feet so that your body is facing the ground in a straight, unrounded perfect plank position (as though you were about to do a push-up). Maintain this position for three minutes, using your stomach muscles to remain in a straight line. You will feel your strong inner woman come alive with this challenging exercise. This truly is an all-over body exercise.

natural beauty

"Ab-Fab" Sit-Ups

Lie on your back with knees bent and feet firmly on the ground. Interlock your hands and place them gently behind your head with your elbows wide. Lift directly up to the sky (this is not your classic forward-bending crunch). Repeat ten times to conclude one rep. Complete three reps.

Simple Stretching

Hamstring stretch: Start by lying on your back. Lean forward with your legs extended and reach for your toes. Indulge in this stretch for as long as your body feels it needs it.

Lower back stretch: Start by lying on your back. Pull your knees into your chest and give yourself a hug. To get a little massage out of this stretch, rock side to side in the stretch position.

Spinal stretch: From a standing position, reach to touch your toes and roll your body back up to a standing position. Do this stretch slowly to articulate the spine. Stretching is for you, so do as many reps as your body enjoys.

6

Common Beauty Ingredients and Helpful Terms

NATURAL INGREDIENTS FOR BEAUTIFUL SKIN AND HAIR

ALMOND OIL (SWEET ALMOND OIL): As a moisturizing ingredient for your skin and hair, almond oil also helps your skin retain its own natural moisture. You can find almond oil in the beauty/body aisle of your local health food store.

ALOE: Aloe heals wounds and soothes sun exposure. It's also a great gentle moisturizer for your skin. Aloe vera is the name of the plant that we get the soothing aloe vera gel from. You can find aloe vera at any drugstore or in the body aisle of your local grocery store.

APPLE: Apples contain natural fruit acids, which help exfoliate the skin and slough off dead skin cells. Apples can be found at your local grocery store or farmers' market.

APPLE CIDER VINEGAR: This is a multipurpose skincare and hair care natural remedy, with clarifying properties, acne-fighting vitamin A, and natural astringent qualities. You can find apple cider vinegar at your local grocery store.

AVOCADO: Avocados are naturally hydrating when added to beauty recipes for both the skin and hair. You can find avocados at your local grocery store or farmers' market.

AVOCADO OIL: This oil improves the texture of your hair and is moisturizing to your tresses. Avocado oil, much like a mashed-up avocado, has hydrating benefits when applied to your skin. You can find avocado oil in the beauty/body aisle at your local health food store.

BAKING SODA: This cleansing agent is useful in removing product buildup in hair and can even be used as a dry shampoo. Baking soda is also a makeup remover to reveal your natural pretty skin. Baking soda can be found at your local grocery store or drugstore.

natural beauty

BEESWAX: Beeswax is known to lock in moisture and protect your skin from damaging environmental factors by creating a barrier between your skin and the outer environment. It's a great moisturizer for both your skin and lips. Beeswax can also be used as a hair beauty recipe to keep your baby hairs, or "fly-aways," in place. You can find beeswax at your local drugstore or in the beauty/body aisle of your local grocery store.

CALENDULA FLOWERS: The calendula flower is associated with skin health. It has antifungal and antimicrobial properties and is an excellent skin purifier. Calendula comes in powder form, or you can extract the oil from the flower's petals with a mortar and pestle. It can be found at your local health food market.

CARROT: Carrots are rich in vitamin A, known for its healing effect on acne. Carrots can be found at your local grocery store or farmers' market.

CAYENNE PEPPER: The strong anti-inflammatory properties make this a surefire way to get the fiery lines out of the whites of your eyes and eliminate bloodshot peepers. This ingredient must be *heavily* diluted with water. Cayenne pepper is found in the spice section of your local grocery store.

CHAMOMILE TEA: This calming drink is known to condition and soften your hair in addition to brightening it. The skin benefits include healing properties for acne and rosacea, as well as exfoliating benefits. It's a great anti-inflammatory ingredient for under-eye puffiness. When applied on hair, it has a natural blonding effect. You can find chamomile tea at your local grocery store.

CHAMPAGNE: Want your blonde hair to become blonder? This blonding agent enhances your natural highlights. Champagne is sold at your local grocery store or wine shop, and you can purchase it for use in beauty recipes if you are of drinking age.

CHLOROPHYLL (LIQUID): Chlorophyll, like vitamin E, is wound healing, but it's also a natural deodorizer. It's found in parsley, which freshens breath. Liquid chlorophyll can be found at your local drugstore or in the body aisle of your local health food store.

CLAY (BENTONITE CLAY, GREEN TEA CLAY, SEA CLAY, FRENCH GREEN CLAY): These detoxifying clays are clarifying agents incorporated

natural beauty

into beauty recipes to draw out toxins and reveal a clear complexion. Clays come in either wet mixed form or dry powder form, ready to be mixed with the appropriate ingredients for a mask. You can find these clays in the beauty/body aisle at your local health food store.

COCOA BUTTER: You can apply this to your skin as an ultimate moisturizer. It creates a barrier between your skin and the environment, locking in moisture. Pregnant women commonly apply cocoa butter to their growing bellies to prevent stretch marks. When applied to your hair, it strengthens the strands, preventing breakage. You can find cocoa butter in the beauty/body aisle of your local grocery store or health food market.

COCONUT OIL: Coconut oil is found in a solid form, which then melts with your body temperature. It's highly moisturizing and helps improve under-eye wrinkles. Coconut oil applied to hair is also hydrating. You can find coconut oil at your local grocery store.

COFFEE GROUNDS (CAFFEINATED): Coffee grounds support healthy circulation and when added to body scrubs can help reduce the appearance of cellulite. Brunettes can keep their color-treated tresses a rich brown shade by mixing coffee grounds with almond oil as a color-supportive beauty recipe for radiant dark hair. Raven-haired beauties can enhance their hair color by brewing coffee, allowing it to cool completely, and using as a rinse before shampooing. Ground coffee can be found at your local grocery store.

CORNSTARCH: A natural ingredient derived from corn, cornstarch assists in the binding of other ingredients. It also prevents skin irritation by providing a protective layer between your skin and the environment. Cornstarch is sold in the baking aisle of your local grocery store.

CUCUMBER: This healing vegetable is widely known to help soothe under-eye puffiness. It's rich in vitamin C, a valuable antioxidant for skin health. The silica in cucumber is helpful for hair growth. You can find cucumbers at your local grocery store or farmers' market.

EGG WHITE: With pore tightening and toning qualities for the skin and shining benefits for your hair, egg whites are an essential part of your natural skincare regime. You can find eggs, the source of this ingredient, at any grocery store.

EGG YOLK: Consisting mostly of fat and water, egg yolks are water binding, leaving your skin hydrated and supple. Egg yolks are also

natural beauty

hydrating when used as a natural ingredient in hair recipes. You can find eggs, the source of this ingredient, at any grocery store.

ESSENTIAL OILS: All essential oils are hydrating for your hair and moisturizing for your skin. Moreover, various types of essential oils have other benefits specific to skin and hair alike. Essential oils can be found at general health food stores or in the beauty/body care aisle of your local grocery store.

Almond essential oil is extremely moisturizing and an excellent source of calcium and vitamin E, which contributes to skin rejuvenation. It's hydrating for your skin, lips, and hair.

Chamomile essential oil is calming and has a pleasant aroma, but it also can help with acne because it removes toxins from your skin. You can certainly add it to a conditioning treatment around your hairline to take in the soothing fragrance.

Eucalyptus essential oil is hydrating for your hair and helps control dandruff. Eucalyptus oil has anti-inflammatory properties for your skin; even very small amounts are cleansing. When used as a fragrance in ingredients, it's refreshing and invigorating.

Grapefruit essential oil has a refreshing aroma and is deeply cleansing.

Lavender essential oil has a calming aroma and enhances your skin's complexion through its antifungal and antiseptic properties, which are healing to acne. It also improves under-eye darkness.

Lemon essential oil has a fresh scent and is antibacterial. It's excellent when used in beauty recipes caring for your feet.

Orange essential oil has a lovely fragrance and is uplifting to the mind. It's antiseptic and has a pleasant flavor should you lick your lips while using a lip balm beauty recipe with orange essential oil in it.

Rose essential oil has astringent and antiseptic properties. It can strengthen your hair and tone your skin.

GINGER JUICE: The juice of ginger is high in antioxidants and gives your skin radiance by minimizing free-radical damage and inflammation. Ginger juice is perfect for brightening skincare beauty recipes. Ginger, the source of this ingredient, can be found at your local grocery store or farmers' market.

GRAM FLOUR: Gram flour is a whitening skin agent and if used periodically as a mask, gives you a bright, radiant complexion. It can be found in the baking aisle of your local grocery store.

GRAPE SEED OIL: Rich in antioxidants and essential fatty oils (including linoleic acid, oleic acid, and palmitoleic acid, all of which are vital to skin health), grape seed oil naturally fights free-radical damage. It's a natural antiaging ingredient for skin care. You can find grape seed oil in the beauty/body aisle at your local grocery store or health food market.

GREEN TEA: High in antioxidants, green tea fights inflammation and can de-puff your swollen eyes. You can apply the steeped tea bags directly to your eyes. Green tea can be found at your local grocery store.

HONEY (NATURAL HONEY, RAW WHITE HONEY, EUCALYPTUS HONEY): Honey is nature's moisturizer with incredible skincare benefits for keeping your skin hydrated. It's antibacterial so it keeps your pores clear from oil and dirt that cause breakouts. It's also a great path to shiny, healthy hair because it packs in the moisture. Honey can be found at your local grocery store or farmers' market. Raw white honey and eucalyptus honey are more rare and are likely to be found at your local health food market.

Jojoba beads (colored variety): Jojoba beads are relatively small beads that are used for gentle exfoliation. They are made from jojoba wax and can be found in the body aisle at your general health food store. You can add them to your bathtub for a soft scrub before you shave your legs.

Jojoba oil: Jojoba oil has high levels of minerals and fatty acids that are important for keeping your skin youthful and radiant. It's hydrating and seals in moisture. Jojoba oil is antimicrobial, therefore cleansing to your pores. You can find jojoba oil at any general health food store or in the beauty/body aisle at the grocery store.

Kelp (powdered): Kelp is seaweed, which is wonderful for radiant skin and hair. It restores moisture in the skin and revitalizes and firms. Kelp adds shine to your hair. The powder form exfoliates the skin, removing impurities from your skin's pores. You can find kelp powder at you local health food store, at Asian markets, or online.

Lavender herb powder: Lavender is a flower; the herb powder is the dried flower of the lavender. All of the same skin benefits of being anti-inflammatory and functioning as a circulatory stimulant are also present in this dry, powdered form of the flower. It can be mixed with hair conditioner to add a lovely fragrance and soften your strands. You can find lavender herb powder from various online sources where aromatherapy products are sold and in the beauty/body aisle of specialty health food stores.

LEMON JUICE: Lemon juice is antibacterial and a natural astringent with toning properties. Lemons can also bring out blonde highlights naturally when squeezed directly onto light-colored hair before spending time in the sunshine. Lemons, the source of this ingredient, are found at your local grocery store or farmers' market.

LIME JUICE: Lime juice is potent in vitamin C, which is known to brighten your complexion. Limes, the source of this ingredient, are found at your local grocery store or farmers' market.

LIQUID MOISTURIZING SOAP: This is the liquid form of a bar of soap, which has antibacterial properties ideal for cleansing in a beauty recipe for your feet. It's a gentle way to cleanse your skin as well. It's too drying as a shampoo, though. You can find liquid moisturizing soap at any drugstore or local grocery store.

MESDEMET: Mesdemet is made from a mixture of copper and galena, also known as lead ore. The women of ancient Egypt used mesdemet to dramatically line their eyes, as Cleopatra made famous.

MILK: Milk is an exceptional source of alpha hydroxy acids, which are naturally exfoliating for your skin. Milk also has protein in it, which is great for strengthening your hair. For clear skin and strong hair, apply milk topically. Milk is found at your local grocery store.

MILK POWDER: We have learned that milk has exfoliating properties, but it also helps to even out skin tone and acts as a skin-brightening agent. Milk in the form of powder is no different than the liquid form, other than that one is wet and one is dry. Milk powder, also called powdered milk, is more convenient to store among your other beauty ingredients, but it still must be refrigerated. You can find powdered milk at your local grocery store.

MINT (LEAVES, OIL): The high levels of antioxidants and flavonoids in mint leaves fight inflammation and reduce that puffy look under your eyes. To seal split ends, apply a teaspoon of mint oil to the ends of your hair, as you would a serum. Mint leaves are an important ingredient for natural beauty and can be found in the fresh herbs

natural beauty

section at your local grocery store or farmers' market. Mint oil can be found in the beauty/body aisle of your local health food store.

OATMEAL (UNCOOKED): The texture of oatmeal lends itself for inclusion in gentle exfoliating scrubs. Oatmeal has skin-soothing benefits and is nourishing for your hair. You can find oatmeal at your local grocery store.

OLIVE OIL (EXTRA-VIRGIN OLIVE OIL): Improving a dry scalp is one way that olive oil supports a great head of hair. It's also hydrating for your strands and your skin. You can replace drying soap with moisturizing olive oil for a silky, hydrating shave. Olive oil of any kind can be found at your local grocery store in the salad dressing/condiments aisle.

PAPAYA: The abundance of alpha hydroxy acids in papaya is naturally exfoliating and helps remove dead skin cells to reveal a brightened, glowing complexion. Papayas can be found at your local grocery store or farmers' market.

PEACH: Peaches are rich in vitamin C, which is important for the body's collagen production. Collagen keeps our skin soft and youthful. Peaches can be found at your grocery store or farmers' market in season. If they are not in season, you can find a bag of frozen peaches in the freezer aisle of the grocery store.

PETROLEUM JELLY: Effective as a moisturizer, petroleum jelly is soothing and hydrating even to very delicate skin. You can also apply it to remove your eye makeup and hydrate your under-eye area at the same time. You can find petroleum jelly at any drugstore or grocery store.

POTATOES: Sliced potatoes are a great puffy eye remedy. They are rich in vitamin B, which is known to reduce inflammation and puffiness. You can find potatoes at your local grocery store or farmers' market.

PURIFIED WATER: Purified water is preferred over regular tap water to ward of bacteria and other impurities.

ROLLED OATS (PLAIN): Rolled oats have a gritty texture that is desirable when your skin is in need of a natural exfoliant that won't be too harsh. It has moisturizing benefits for your skin as well. You can find rolled oats at your local grocery store.

Rose: A rose is a thing of beauty. Roses smell divine and offer skin and hair care benefits in abundance. You can add the petals of the actual flower to your bath, or you can extract the oils from the petals with a mortar and pestle to add as a fragranced detangler to damp, washed hair. You can find roses at your local flower shop or from an online source that delivers flowers. If you live in an urban area, there might be a flower district where you could buy them in bulk at a discounted price.

Rosewater: Rosewater cleanses the skin with its clarifying properties. It removes dirt, oil, and other impurities from within the skin's pores. It's also a natural toner. If you saturate a cotton ball with some rosewater after you have cleansed your face, it will tone and balance your skin. You can add rosewater to your shampoo for a beautiful fragrance and for its moisturizing properties. You can find rosewater in the beauty/body aisle of your local health food store.

Sandalwood powder: Sandalwood powder is a natural way to improve the appearance of scars. It also serves to soften your skin in the process of healing your scars. You can find sandalwood powder at your local health food markets, at incense stores, or online.

SEA SALT: Sea salt is an excellent natural exfoliant for your skin. You can grind coarse sea salt with a mortar and pestle or purchase it as sea salt in the baking aisle from your local grocery store.

SHEA BUTTER: Packed with vitamins A and E, shea butter is moisturizing and penetrates your skin deeply to restore collagen and elasticity. You can use shea butter in any number of moisturizing beauty recipes as a base, and you can apply it directly to the ends of your hair for conditioned locks. Shea butter can be found in the beauty/body aisle at your local health food market.

SILICA: Silica aids with hair growth and is found in cucumbers. Cucumbers, the source of this ingredient, are found at your local market or farmers' market.

SOUR CREAM: Sour cream has a high value of lactic acid, which functions as a gentle exfoliant on your skin, removing dead skin cells without the granular texture or rubbing. It reveals a clearer complexion. Putting sour cream on your damp hair seals the cuticle, preventing split ends.

SUGAR (WHITE OR BROWN): Sugar is a grainy substance that is excellent as a natural exfoliant. It's great for body scrubs and facial exfoliating treatments. Brown sugar can remove excess oils and impurities from your scalp when mixed with equal parts olive oil for a mild scrub to reveal clean, radiant hair. Sugar can be found in the baking section at your local grocery store.

TANNIN: Tannin is a bitter plant compound with astringent properties. It is found in witch hazel. Witch hazel, the source of this ingredient, can be found at any drugstore.

TURMERIC POWDER: This special powder is antiaging, anti-inflammatory, and antibacterial. It's used in beauty recipes to even out skin tone. Turmeric powder can help reduce skin rashes as well. It can be found in the spice section of your local grocery store.

VITAMIN C CAPSULES: Vitamin C aids collagen synthesis, maintaining the skin's elasticity and improving overall texture. You can find vitamin C capsules at your local health food market.

VITAMIN E OIL: This is intensely moisture retaining and hydrating when applied as part of a hair-conditioning beauty recipe or as part of your natural skincare routine for drier skin. Vitamin E oil can be found in the beauty/body aisle of your local health food store.

WITCH HAZEL: Witch hazel is a wonderfully effective astringent with toning properties. It's also moisturizing and helps skin maintain its natural moisture and elasticity. Witch hazel can be found at any drugstore.

YOGURT (PLAIN): Packed with minerals such as zinc, calcium, and B vitamins, yogurt helps reduce pore size and improve skin texture. As a hair treatment, yogurt is softening. You can find yogurt at your local grocery store.

NATURAL SKINCARE TERMS

ALPHA HYDROXY ACIDS: Alpha hydroxy acids occur naturally in fruit, milk, and yogurt (among other foods) and are used in beauty recipes for the naturally exfoliating effect they have on the surface layer of the skin.

ANTIMICROBIAL: This substance destroys and reduces the growth of disease-causing microorganisms.

ANTIOXIDANTS: These substances inhibit the process of aging by fighting free-radical damage, also known as oxidation.

natural beauty

AROMATHERAPY: This includes the use of any fragranced essential oils, scented candles, or fragrances natural to certain ingredients that are added to beauty recipes to induce a desired state of mind or mood such as relaxation, calm, or invigoration.

BETA HYDROXY ACIDS: Beta hydroxy acids accelerate skin cell turnover and help clear pores. They occur naturally in some fruits such as strawberries and in milk.

BLACKHEADS: A blackhead is known as an open comedone (see following) because the sebaceous glands (oil-producing glands) are clogged with dirt and dead skin cells, giving the black appearance of the head at the opening of the hair follicle.

COMEDONE: This is another term for a blemish or pimple (see "Blackhead" and "Whitehead").

DÉCOLLETAGE: This is the lower part of a woman's neck and upper part of her breasts not covered by a low-cut dress.

DE-PUFF: This term is used to describe a reduction in the puffy appearance of the under-eye area.

DETOXIFY: This means to clarify your skin's pores and eliminate toxins through the skin.

EMULSIFY: To emulsify is to blend an emulsion, or mixture, until all ingredients are entirely mixed.

EXFOLIATE: This means the removal of oil, dirt, and dead skin cells from the surface of the skin either via ingredients alone or via a gentle scrubbing motion of the beauty recipes on the skin.

HOMEOPATHY: This term refers to a natural remedy system for treating skin or health issues. Certain ingredients found in these beauty recipes, such as calendula, are also known to be part of homeopathic healing treatments. It's a natural, holistic approach to wellness.

HYPERPIGMENTATION: This refers to the darkened discoloration of certain areas of your skin, commonly occurring as a result of sun damage. For this reason, hyperpigmentation is also referred to as sun spots.

INFUSE: This term is generally used when talking about steeping tea in natural beauty recipes. It's the slow mergence of a liquid into another liquid.

natural beauty

LATHER: This means to create a frothy, sudsy substance by mixing an ingredient or ingredients with water.

PURÉE: This means to vigorously blend beauty ingredients until they become liquid.

RESIDUE: Residue refers to a buildup of dirt, oil, or product on the scalp, hair, or surface of your skin.

SERUM: This is a liquidlike, silky-textured substance created using natural beauty ingredients for either skincare or hair care.

VASOCONSTRICTION: This is the process of blood vessels constricting, which often results in reducing inflammation or puffiness.

WHITEHEAD: A whitehead is a pimple, or closed comedone, defined as sebaceous glands clogged with oil and dead skin cells in the opening of a hair follicle.

NATURAL BEAUTY APPLICATORS
AND NECESSITIES

The following is a list of applicators you will need for some of the beauty recipes in the chapter to follow:

- Foundation brush or paint brush with synthetic hairs/fibers
- Natural loofa sponge or sea sponge
- Cotton balls or cotton pads
- Cotton swabs
- Tongue depressors
- Eyedroppers
- Mixing bowls, small and medium (microwave safe)
- Airtight glass storage containers, various small sizes
- Misters or empty bottles
- Washcloth (for face)
- Shower cap (plastic)
- Hair clip, large enough to hold all of your hair
- Medium-sized pot for boiling, kitchen item
- Strainer, kitchen item
- Blender, kitchen item

7

*

Beauty Recipes

Natural beauty recipes are essentially affordable luxuries. You deserve to reach for beauty just as much as the women who came before you and the women of future generations who will reach for their highest beauty. My beauty recipes are a natural way for women to indulge in the inevitable desire to reach for beauty.

Lustrous Locks

If you've been spending time at the beach, you may have noticed your hair looking a bit dry. This is an easy fix and should be indulged in bimonthly during the summer months.

1 egg yolk
1 tablespoon almond oil
1 cup water

Preparation and Application
In a small bowl, mix ingredients until emulsified, and apply to your hair as a conditioning treatment in the shower. Rinse with warm water.

Try not to blow-dry your hair too often because that also dries hair out. Take advantage of the season's sunshine and let your hair air-dry.

Apple Cider Rinse

If your hair has a buildup of product, or is just tired looking from environmental residue and oils, a quick beauty recipe for shiny hair is just what you need. It's so simple.

1 cup apple cider vinegar

Preparation and Application
Before shampooing your hair, pour the apple cider vinegar on your head in the shower, with your head tilted back. Gently work through your scalp and hair, as if you were shampooing, although the vinegar won't produce any suds. Rinse with the warm stream of water from your showerhead and follow with a shampoo and conditioner, as usual. The shampoo will remedy the vinegar smell and will get incredibly sudsy because of the apple cider vinegar stripping the dirt and oils away for ultraclean scalp and hair.

Egg Protein Mask

Egg whites have tremendous protein benefits. We want to nourish our hair with an intensive egg-white protein mask for shiny and healthy stands of hair.

> 3 egg whites
> 5–7 drops lemon essential oil

Preparation and Application
Crack three eggs over a bowl, separating the whites (acne sufferers, save the yolks and for the Yolk Facial recipe on page 98). Once the egg whites are isolated in a bowl, whisk together until fluffy. Add the drops of lemon essential oil for a pleasing aroma. Use your fingers to apply the mask to damp hair, focusing on the ends where the hair naturally weakens. Set hair in a clip and place a shower cap over masked hair while showering. Before exiting the shower, thoroughly rinse out the protein mask with cold water. Comb through hair gently and let air-dry.

Olive Oil Scalp Massage

This beauty recipe is all you need to improve a dry and irritated scalp.

    3 tablespoons olive oil

Preparation and Application

Pour olive oil into a small bowl. Take the bowl into the shower with you (use a plastic bowl for safety). Wet your hair and separate your hair into parts. Pour the olive oil directly on your scalp at each part of your hair. Massage gently for a couple of minutes before shampooing and conditioning as usual.

Try not to shampoo your hair every day; that dries out your scalp and your hair. Instead, wash your hair twice weekly for optimum scalp rejuvenation.

Champagne Locks

If you covet European blonde hair, then you should mimic the French and add champagne to your hair care beauty regime. Champagne Locks is the perfect recipe for shiny, soft, champagne-highlighted hair. Chamomile teas, just as champagne and lemons do, have a blonding effect on your strands and enhance your natural highlights. Chamomile also has healing properties to condition and soften your hair. Celebrate your tresses with bubbly!

> 1 cup (2 tea bags) chamomile tea, steeped
>   and cooled
> Juice of 1 lemon, squeezed
> 1 cup champagne

Preparation and Application
Steep two bags of chamomile tea in 1 cup of boiling water, covered, in a medium-sized pot. Allow tea bags to infuse the water. Remove tea bags and allow water to return to room temperature. Squeeze in the juice of one lemon (don't worry about the seeds or pulp getting in

natural beauty

the pot). Add 1 cup of champagne and stir. Pour through a strainer into a bowl. Take into the shower and drench your hair with this beauty recipe after shampooing. Let the sunshine dry your hair to reinforce the highlights. Shine on!

Healthy Hair

Every woman wants healthy, beautiful hair, right? In the days of bleaching, hair spraying, and harsh styling products, healthy hair is becoming harder to achieve. This beauty recipe will support you in the fight for healthy hair, so you don't have to give up the styling routines you use to reach for beauty.

    5 drops rose essential oil
    5 drops lavender essential oil

Preparation and Application
Mix oils together and place into the palms of your hands and rub together. Coat the ends of your damp hair. Brushing your tresses distributes the natural oils from the scalp to the ends of your strands of hair, where the moisture is needed most. Leave in to seal the ends and style as normal.

    It's always a great idea to allow your hair to dry naturally because the intense heat from a blow-dryer is damaging and splits the very ends of the hair that this beauty recipe is meant to heal.

Cucumber Conditioner

Silica is rich in potassium, which is great for strong hair and nails. Cucumber has high levels of silica; therefore, it's beneficial for promoting hair growth. Cucumber Conditioner is an excellent natural beauty recipe that you can also concentrate on your scalp area to stimulate the hair follicles.

¼ cucumber, peeled
1 egg
4 tablespoons olive oil

Preparation and Application
Place cucumber in the blender until puréed. In a small bowl, mix the egg, olive oil, and cucumber until smooth. Spread evenly through your hair, including the crown area and scalp. Allow to sit for 5 to 10 minutes while in the shower. Rinse well. This is a good time to shave your legs while your hair is being deeply conditioned.

Baking Soda Savior

Rid your hair of residue commonly associated with product buildup.

1 tablespoon baking soda
2 tablespoons of your favorite shampoo

Preparation and Application
Mesh the baking soda with your shampoo. Shampoo your hair as you normally do. The addition of the baking soda makes for a wonderful lather. This beauty recipe clarifies your hair.

natural beauty

Avocado Oil Enhanced Conditioning

The best part of this beauty recipe is that it creates shiny strands of hair because of the intensive moisture retention capabilities of the avocado oil and vitamin E oil.

    1 cup avocado oil
    3–4 drops natural liquid vitamin E oil
    2 tablespoons of your favorite conditioner

Preparation and Application
Add the avocado oil and vitamin E oil to your favorite conditioner for an enhanced nourishing treatment. Massage into clean, shampooed hair, focusing on the ends, which tend to be the driest.

The vitamin E protects your hair and scalp. The fatty acids in avocado oil nourish your hair for shiny locks.

Beauty Byte

KEEP STRANDS SHINY AND HEALTHY

→ Indulge in regular oil treatments, and apply a couple drops of oil on damp hair as a serum.

→ Shampoo only once a week. Other times just wet your hair and apply conditioner.

→ Trim or dust the end of your hair every couple of months, even if you are growing out your hair. Split ends can't become unsplit.

→ Let your hair air-dry whenever possible to spare your hair the heat from your hair dryer.

natural beauty

Hot Oil Treatment

Ladies, what is more important than sexy hair? Enjoy this beauty treatment to get your dry hair back on track. You will have some time while your hair is getting the shine treatment. I like to paint my nails while I wait for my hair to become bombshell locks!

1 cup extra-virgin olive oil
3–4 drops eucalyptus essential oil

Preparation and Application
Mix oils together and place in a small microwave-safe bowl (I favor a small glass bowl). Microwave for 30 seconds. Massage hot oils into dry, damaged hair focusing on the ends, which are usually the most damaged. Cover your hair with a plastic shower cap for 20 to 25 minutes. Then rinse thoroughly. Wash with a delicate shampoo.

Soothing Cucumber

Cucumber is well known for having a soothing effect and benefiting under-eye puffiness. This is thanks to the acid and foliates in cucumber. This recipe takes beautiful eyes to the next level.

¼ cucumber
1 tablespoon aloe
¼ tablespoon cornstarch
1 tablespoon witch hazel

Preparation and Application
Place cucumber in the blender and purée it. Mix the fresh cucumber juice with aloe and cornstarch. Heat the mixture in a small microwave-safe bowl for 45 seconds. Remove from the microwave and, using a tongue depressor, stir in the witch hazel. Allow the mixture to cool (it will become creamy and clear). Gently apply under your eyes and reap the benefits of cucumber to tighten the bags underneath your tired eyes. Leave on for 15 minutes. Then gently rinse with warm water.

natural beauty

Smooth as Butter

Petroleum jelly is an effective eye makeup remover, even for waterproof mascara. But I like to use it in eye cream for its moisturizing properties. Cocoa butter is exceptionally high in vitamin E, which helps to soothe and hydrate the delicate skin under your eyes. Coconut oil is also extremely hydrating and helps improve under-eye wrinkles.

> 1 tablespoon petroleum jelly
> 1 tablespoon cocoa butter
> 1 tablespoon coconut oil

Preparation and Application

Combine all ingredients in a microwave-safe bowl and heat in the microwave for only 10 seconds. Cocoa butter and coconut oil melt and absorb easily with the help of your body heat, so no more than 10 second in the microwave is necessary. Pour melted mixture into a clean travel-size jar and let cool. Use as a nightly under-eye moisturizer. Your eyes will be smooth as butter.

Keeping the delicate skin under your eyes hydrated is the secret to preventing fine lines and wrinkles from developing or worsening.

Calming Chamomile Eye Serum

For this recipe you will need an eyedropper, available in the beauty or travel sections of most grocery stores. You may even find them at some drugstores.

> 3 tablespoons jojoba oil
> 4–5 drops lavender essential oil
> 4–5 drops chamomile essential oil

Preparation and Application
In a small bowl, mix oils together. Shake well before each use. Apply one drop per eye of this beauty serum to your under-eye area before bedtime and in the morning on a clean face.

Chamomile essential oil has a pleasing and relaxing aroma, as does the lavender oil. And lavender has healing properties that are perfect for improving dark under-eye circles.

natural beauty

Beauty Byte

TIPS FOR BRIGHT EYES

→ Use my Cooling Mint recipe on page 86 to de-puff tired eyes.

→ Get sleep, real sleep, not just restless time spent on your pillow. Reading before bed is the best way to induce a peaceful slumber.

→ If you opt for no makeup, curl your eyelashes with an eyelash curler. It will open up your eyes in a big way.

→ Your eyebrows frame your eyes, so shape them! Use whatever preferred method of reaching for beautiful brows you prefer—waxing, tweezing, threading, or even laser hair removal. Anyway you reach for beauty is perfect, as long as your brows are shapely and frame your pretty eyes nicely.

→ Keep eyes clear with eye drops if your eyes tend to get irritated.

→ Vanish under-eye darkness with an under-eye concealer to match your skin tone.

Bloodshot Be Gone

Bloodshot eyes can reveal themselves from fatigue, sleep deprivation, or stress. But it's precisely a lack of oxygen to the cornea, as the small blood vessels become inflamed and produce a bloodshot appearance. A great way to fight this look is with cayenne pepper, which has strong anti-inflammatory properties that will shrink you blood vessels and avoid the inflammatory bloodshot appearance.

    Pinch cayenne pepper
    1 cup water
    2 cotton balls

Preparation and Application
Take a tiny pinch of cayenne pepper and heavily dilute it with the water. Once diluted, place the liquid on a cotton ball or soft tissue and gently blot over closed eyes. Splash face and eyes with warm water to avoid irritation. *Be especially cautious not to get the cayenne pepper into your actual eye. It is quite painful.*

Wondrous Witch Hazel

Witch hazel helps combat dark under-eye circles. It's a remarkable anti-inflammatory agent and a highly effective astringent. Witch hazel contains high levels of tannin, which is a bitter plant compound with astringent properties.

5–6 drops witch hazel
1½ cups water

Preparation and Application
Add the drops of witch hazel to the small bowl of water, thereby heavily diluting the witch hazel. Allow to cool overnight in the refrigerator (you want the liquid to be cool and soothing). Saturate two cotton balls with the witch hazel solution and gently apply to closed eyes. Splash some cool water on your face and eye area afterward. You will feel refreshed and will be improving your dark under-eye circles at the same time.

Potatoes for Puffiness

Believe it or not, sliced potatoes can de-puff your swollen eyes. Bags, or under-eye puffiness, are caused by a number of variables, including a diet high in sodium, alcohol consumption, fatigue, or stress. Potatoes are packed with B vitamins known to reduce inflammation and puffiness.

2 potatoes, thinly sliced

Preparation and Application
Place the round root of the potato over your entire eye, leaving it on for 10 minutes. Rinse off the residue, or potato film, with cool water.

natural beauty

Green Tea to the Rescue

Green tea is a powerful antioxidant, which helps combat inflammation caused by free radicals in the environment. Oxidants, or free radicals, wreak havoc on our skin and can contribute to darkness under the eyes.

    2 green tea bags

Preparation and Application
Steep two green tea bags and set aside to cool. Once cool, blot them a bit and hold under your eyes.

    Beyond antioxidant support for dark circles, the caffeine in green tea is excellent for a process of blood vessel constriction, or vasoconstriction, which reduces puffiness.

Milk-Soaked Cotton

Milk is an exceptional source of alpha hydroxy acids, responsible for exfoliation. This beauty recipe is a unique way to gently exfoliate the sensitive skin under your eyes, removing the upper layer of skin cells and allowing new healthier cells to take their place for bright eyes.

½ cup whole milk
2 cotton balls

Preparation and Application
Saturate two cotton balls with the whole milk. Gently dab the milk-soaked cotton balls onto the skin under your eyes and allow the milk to absorb and dry under your eyes for 15 minutes before rinsing away the exfoliating milky residue with warm water. Your eyes should feel tighter and renewed.

This not only feels soothing but also reduces the appearance of dark circles.

Peachy Peepers

Fresh peaches are rich in vitamin C, which is important for the body's collagen production. If your eyes have a sunken or shadowy look, peaches are just the thing to bring your under-eye skin back to its supple and youthful bright-eyed look. Yogurt is thick, creamy, and packed with B vitamins and minerals such as zinc and calcium. It also helps to reduce pore size and improve skin texture. The skin under your eyes is delicate, and this is the perfect way to treat your tired eyes.

¼ ripened peach
1 teaspoon plain yogurt

Preparation and Application
Mash the peach in a bowl with a fork or in a mortar with a pestle. Add the yogurt and mash together. With your fingers, apply the creamy textured mixture under your eyes. Allow the peach to work its nourishing, anti-inflammatory magic for 10 to 15 minutes and rinse gently with warm water.

Cooling Mint

Fresh mint leaves are packed with flavonoids and antioxidants, which fight inflammation and reduce that puffy look under your eyes. Sweet almond oil has a plentiful supply of vitamins E and D, with high levels of fat. It softens and repairs the delicate skin under your eyes, reducing the appearance of dark circles.

> Small handful mint leaves
> 10 drops sweet almond oil

Preparation and Application
Wash your batch of fresh mint leaves and pat dry with a paper towel. Take a small handful of the leaves and grind in a mortar and pestle with almond oil until you have a fine oily substance. Gently apply to the delicate skin under your eyes, avoiding getting the mixture into your eyes, and leave the mint oil on for 15 minutes before rinsing with warm water. Your eyes should feel cool and refreshed.

natural beauty

Almond Hydrated Eyes

Almonds are extraordinary seeds given to us by their trees in nature. The almond essential oil is hydrating and helps the skin maintain its own moisture. Rich in vitamins E and D, almond oil has reparative qualities for your skin. It's also high in antioxidant value, deeming it anti-inflammatory and making it a gentle moisturizer ingredient. Coconut oil is immensely moisturizing. This combination is great for tired eyes.

5–6 drops almond essential oil
1 tablespoon coconut oil

Preparation and Application
Mix ingredients together in a small bowl. Use a cotton swab to smear the mixture under your eyes. Once it has been deposited on your skin, use your finger to gently rub over the entire under-eye area. The substance will be slippery as the coconut oil melts with the temperature of your body (this is why you need only a small dab with a cotton swab to apply this moisturizing recipe). Allow

… continued on next page

to absorb into your under-eye skin for 10 minutes. Rinse thoroughly with warm water.

This is highly hydrating, and you will immediately notice a difference in the quality of your skin. Your concealer will go on smoother, and if clumping was a previous problem because of dry skin, consider that problem solved.

Soft Kisses

Soft lips are desirable, and—let's be honest—dry lips are uncomfortable. A perfect lip moisturizing gloss is always a must. Let's make our own! Aloe vera is naturally soothing, and both petroleum jelly and coconut oil serve to retain moisture and hydrate.

1 teaspoon aloe vera
1 teaspoon petroleum jelly
½ teaspoon coconut oil

Preparation and Application

In a small microwave-safe bowl, mix ingredients until blended smoothly. Place bowl of ingredients in the microwave and heat for 1 minute or longer if necessary (the mixture should become a liquid). Stir and pour into a small glass or plastic travel jar for personal use once it has cooled down.

Pucker Up

Chapped or cracked lips can be uncomfortable and even pain-ful. Making sure to stay hydrated by drinking a lot of water is important during the hot summer months and the dry winter months. But you can also make this exfoliating lip scrub to smooth out your lips so you're ready to pucker up.

    1 teaspoon sea salt
    1 teaspoon petroleum jelly

Preparation and Application
Combine ingredients and gently rub over your lips as though you were applying a ton of lip gloss. Rinse clean with warm water.

    I recommend applying an SPF lip balm year round to prevent peeling lips. Blistex is excellent and has an SPF value of 15 for protected, kissable lips.

Luscious Lips

Pucker up with soft, supple lips. The summer heat can be harsh and cause chapped lips or even sun blisters.

½ cucumber
1 tablespoon honey
1 tablespoon plain yogurt

Preparation and Application
Peel the cucumber and purée it in a blender to extract the juice. Add the honey and plain yogurt. Stir until your homemade lip balm is blended. Apply like lip gloss and allow to absorb.

Glossy Lips

Lip gloss is sexy in the summertime, but if your lips are painfully chapped, it will only amplify that painful look. Petroleum jelly and water are your new best friends. This is a straightforward beauty treatment to achieve kissable lips.

Dab of cocoa butter
Dab of petroleum jelly

Preparation and Application

Before you go to bed, drink a glass of water. Then apply your lip-beauty recipe mixture of petroleum jelly and cocoa butter to your chapped lips. After a few days of hydration, inside and out, you will have luscious lips ready for a summer romance!

natural beauty

Sweet Almond Lip Balm

Sweet almond oil is an excellent moisturizer and emollient. These are important characteristics to apply to the state of your lips. We all want supple, soft lips, so let's not be stingy with the sweet almond oil. For this lasting beauty recipe, you'll need a small jar to store your lip balm. You can find such containers in the travel section at most drugstores and in the body section of health food markets.

> 4 ounces sweet almond oil
> 1 teaspoon pink jojoba beads
> ½ ounce beeswax
> 2 teaspoons chamomile essential oil

Preparation and Application

In a small microwave-safe bowl, combine the sweet almond oil, pink jojoba beads, and beeswax. Melt the ingredients for 30 seconds in the microwave. Stir with a tongue depressor. Then add chamomile essential oil and let cool for a few minutes. Pour into your lip balm jar and keep in your medicine cabinet to moisturize your lips daily or nightly.

Sexy secret: Jojoba beads come in a variety of colors, so you can pick a bold raspberry, hot pink, or a more subtle hue. If you are a rock star, yes, they come in aqua blue!

Beauty Byte

HONOR YOUR NATURAL BEAUTY

→ Do with your hair what pleases you aesthetically. If you like to wear your hair straight, then do so. If you like your hair best in beachy waves, then you should wear beachy waves. Your hair frames your face, and the style should appeal to you so that you feel good about yourself when you look in the mirror.

→ If you dread the treadmill, then don't do it! There are other ways to get your bottom in tip-top shape. Go on a hike or break a sweat in yoga. Find an activity that makes you happy. I love yoga. I encourage everyone to find a physical outlet they love as a way to honor their body.

→ Cleanliness is next to godliness. Perfume ought to be sprayed on a freshly bathed body.

→ Always remember to remove your makeup before you go to bed. This keeps your pores clean and reveals glowing skin in the morning. It's proper hygiene, which is an important way of valuing yourself. Take care of yourself, inside and out. Let your natural beauty shine!

natural beauty

For the Love of Olive Oil

Olive oil is a wonderful, natural source of hydration for your skin, internally as well as topically. Our lips crave moisture as we spend our days talking, out in the elements, or wearing lipstick without sunscreen. Even licking our lips occasionally, which is as natural as blinking, can dry them out. This is where olive oil comes in and why you will love it.

4 tablespoons olive oil

1 tablespoon beeswax

4–5 drops orange essential oil

Preparation and Application

In a small microwave-safe bowl, combine olive oil and beeswax. You can grate the beeswax with a cheese grater so that it melts faster if you like. Once it has melted, add the orange essential oil and stir to keep smoothly blended. Pour into a glass travel jar. Let cool and harden before applying it. Apply like lip gloss.

Sugar Lips

Lips are not immune to the accumulation of dead skin cells, and we often exfoliate our skin but avoid the lip area. This is a mistake. Your kisser needs some sugar!

½ teaspoon sugar
½ teaspoon honey

Preparation and Application
Mix sugar and honey together with your fingertips until you have a grainy paste. Apply to lips like a lip gloss, gently scrubbing in a circular motion in an effort to remove the dead skin cells and reveal soft, kissable lips.

Clarifying Clay

The summer heat can intensify oily skin and make your complexion look greasy. Even ordinarily dry skin might have some oil or dirt-induced blackheads. This deeply clarifying mask is your saving grace to reveal a clear face. Bentonite clay is seriously detoxifying and draws the toxins out of your pores, leaving you with toned and tightened clear skin.

> 1 tablespoon bentonite clay powder
> 2 tablespoons purified water

Preparation and Application
Simply add the bentonite clay to the purified water. Stir the ingredients with a tongue depressor until you have a smooth paste. Apply to your face, avoiding the eye area. Leave on for 15 minutes, until it dries and feels slightly tight. Rinse off with warm water and splash cold water on your face afterward to seal the pores.

Yolk Facial

Egg yolks contain high levels of vitamin A, which is a key vitamin for calming down acne.

2 eggs, yolks only
2 tablespoons purified water

Preparation and Application
Over the sink, crack the eggs, and separate the yolks from the egg whites by passing the yolk back and forth between the two half shells. Allow the egg whites to fall into a bowl in the sink and you will be left with the precious yolk. Save the bowl of egg whites for my Egg Protein Mask (page 66) for radiant hair! Add the purified water to your yolk and whip into a froth. Use a tongue depressor to apply to your entire face, avoiding the eye area. Perhaps you have only a couple of acne spots that appear inflamed with a comedone or whitehead. You still want to apply over your entire face because the yolks have the added benefit of reducing redness and returning an overall monochromatic look to your skin.

Lavender Toning Mist

Lavender oil is healing for acne-prone skin. It's also refreshing as a mister with a delightful aroma.

½ cup purified water
5–7 drops lavender essential oil

Preparation and Application
Fill a mister or spray bottle with the purified water, and add the lavender essential oil. You have just made a great summer misting toner with moisturizing properties. Generously mist your face after your typical cleansing routine.

Green Tea Mask

Green tea is beneficial to our skin for its antioxidant, antibacterial, and anti-inflammatory properties. Vitamin C is critical for collagen synthesis, maintaining the skin's elasticity and improving overall texture. Vitamin C capsules can be found at your local health food market.

2 green tea bags
2 tablespoons green tea clay
3 vitamin C capsules

Preparation and Application
In a tea kettle filled with water, steep the bags of green tea until the kettle whistles. Pour ½ cup of the steeped green tea into a small bowl. Add green tea clay and stir with a tongue depressor until you have a smooth paste. Lastly, open the vitamin C capsules and pour into the mixture. Blend. Apply to skin with your fingertips, avoiding your eye area. Allow mask to dry until you feel a slight tightness or pulling. Rinse thoroughly with warm water.

Enjoy the skin-brightening benefits and the layer of protection from environmental toxins that you have just armed your skin with!

natural beauty

Beauty Byte

TAKE YOUR VITAMINS

⚹ It's important to get your nutrients and vitamins from the natural source: food. But cover all of your bases with a daily vitamin supplement. Optimize your beauty—hair, skin, and body—with vitamins and minerals.

⚹ Allow minerals and vitamins to absorb into your body naturally with the topical body beauty recipes in this chapter.

⚹ Drink fresh pressed vegetables and juices to get your daily dose of vitamins and nutrients. You will have an instant energetic response to this natural goodness of liquefied vitamins and minerals.

Honey Healing Clay

This is a unique multipurpose mask. The honey is an incredible moisturizer, and the bentonite clay is a detoxifier.

1 tablespoon bentonite clay
1 tablespoon apple cider vinegar
1 tablespoon raw white honey

Preparation and Application
Mix bentonite clay with apple cider vinegar until you have a thin paste. Add the raw white honey to this supremely luxurious mask. Apply to your face, avoiding the eye area. Allow to dry on your skin for 15 minutes. Rinse gently with warm water. Your skin may have a pinking tint for up to an hour after you have rinsed off this mask. This is normal and will fade, revealing deeply cleansed, taut skin.

natural beauty

Chamomile Steam Facial

Chamomile essential oil has antiseptic and antibacterial qualities in addition to being a wonderful calming source of aromatherapy. This combination is great for a cleansing steam facial. As you open your pores with the steam, the healing chamomile essence permeates your pores and breathes health into your skin.

 2 cups water
 7–8 drops chamomile essential oil

Preparation and Application
Over the stove, bring a medium pot of 2 cups water to high heat, but not quite boiling. Add the chamomile essential oil for a calming aroma. Use a hair band to pull your hair back, away from your face, and lean over the pot. Close your eyes and cover your head completely with a towel as you lean over the steaming pot and allow the natural steam facial to open your pores. Inhale the calming aroma with deep, relaxing breaths until your skin is covered in a wet sheath. With a cool, damp washcloth, pat your face and seal the pores. This chamomile steam facial hydrates, cleanses, and calms your skin.

Gentle Exfoliant

Blackheads are an accumulation of dirt that has been sitting on the surface of your skin and has clogged the pores. It's important to use a facial cleanser first thing in the morning and again before bedtime. Sour cream has a high value of lactic acid, which functions as a gentle exfoliant on your skin, removing dead skin cells as you cleanse your face.

1 teaspoon sweet almond oil
2 teaspoons coconut oil
1 tablespoon sour cream

Preparation and Application

Mix oil ingredients with sour cream until whipped together. Gently apply to your face, avoiding the eye area. Rinse thoroughly with warm water.

natural beauty

Fruit Acid Exfoliant

Apples contain natural fruit acids, which help exfoliate the skin and slough off dead skin cells. Honey is nature's moisturizer and helps the skin retain its natural moisture, in addition to adding hydration. Eucalyptus honey, specifically, is a highly effective antiseptic with antibacterial properties. These aspects of eucalyptus honey are valuable for a healthy, glowing complexion in conjunction with their medicinal qualities. The anti-inflammatory eucalyptus honey has a pleasing aroma that provides a cooling and refreshing effect and induces mental calm.

1 organic apple
1 tablespoon eucalyptus honey (or any honey you
    have available)

Preparation and Application
Core and quarter the apple using a knife and cutting board. Discard all but the seedless slices of apple. Place apple slices into a blender and grind. Pour the ground apple into a small bowl and add the honey. Stir the mixture with a tongue depressor until it becomes slightly silky, but just a bit grainy as well. Apply to a clean face and leave on for 15 to 20 minutes. Rinse with warm water followed by a splash of cool water to seal the pores.

Papaya Milk Mask

The abundance of alpha hydroxy acids in papaya is exfoliating and helps remove dead skin cells to reveal a brightened, glowing complexion. Milk also has the coveted acids, and the combination of papaya and milk is the most gentle and natural exfoliant. Honey is nature's moisturizer, and this nicely balanced mask provides a subtle exfoliation with hydration.

½ papaya
1 tablespoon whole milk
1 tablespoon honey

Preparation and Application

Prepare papaya by slicing the fruit lengthwise and scooping the firm black seeds out with a spoon. Discard the seeds, and scoop the meat of the papaya into a small bowl. Add honey and milk and mash until blended smoothly. Apply to clean face, avoiding your eye area, and leave the mask on for 15 minutes. Rinse with warm water and splash some cold water afterward to seal your pores.

This mask should feel really cleansing, and your skin should be soft to the touch afterward.

Seaweed to the Rescue

Seaweed has a naturally anti-inflammatory effect on acne. It also acts as an exfoliant, removing dead skin cells from the surface layer of your skin. Seaweed, or kelp, in all of its forms is wonderfully clarifying. Rich in B vitamins, seaweed also calms skin irritations and rosacea. Honey is deeply nourishing as a natural moisturizer, so this beauty recipe gives you the benefits of clarifying your skin and providing hydration at the same time.

1 teaspoon powdered organic kelp
1 teaspoon honey
3 tablespoons plain yogurt

Preparation and Application
Blend all ingredients in a small bowl until smooth. Apply mask to face and décolletage and leave on for 10 minutes. Remove with a warm, damp washcloth. Splash cool water on your face afterward to seal your pores.

Brightening Mask

Typically, women alternate between a hydrating mask and a clarifying mask, according to their skin's needs. A brightening mask is a valuable addition to your rotation. A dull complexion can be corrected by using this beauty recipe, which will purify your pores and remove dead skin cells so that new cells can form. Lime juice is high in vitamin C, which is known to brighten your complexion. Your face will feel shiny and new after this beauty mask.

1½ teaspoons gram flour
Pinch turmeric powder
1 teaspoon olive oil
1 teaspoon whole milk
⅓ teaspoon fresh lime juice

Preparation and Application
Mix ingredients together in a small bowl. Apply to face and leave on for 15 minutes. Rinse thoroughly with warm water.

natural beauty

Honey Almond Moisturizer

Almond oil helps your skin retain its natural moisture in addition to boosting hydration when applied topically. Honey is nature's sweetest moisturizer, and your skin will thank you for quenching its thirst!

    2 tablespoons honey
    1 teaspoon sweet almond oil

Preparation and Application
Mix the honey and almond oil. Apply a thin layer to your face, avoiding the eye area. Let nature's ingredients nourish your dry skin for 15 to 20 minutes as a deeply moisturizing mask. Rinse thoroughly with warm water and gently pat skin dry with a washcloth.

Creamy Facial Cleanser

Egg whites have toning qualities that work to tighten your skin and keep your complexion youthful and glowing.

  1 egg white
  2 tablespoons sour cream
  1 tablespoon purified water

Preparation and Application
Mix ingredients together until creamy. Scoop up the mixture with your fingertips and work into a lather to gently cleanse your face daily. Rinse with warm water, and splash your skin with cold water to seal the pores.

Milky Ice Cubes

Milk has alpha hydroxy acids, which are clarifying and serve to tone and tighten the skin. The beta hydroxy acids are exfoliating, so by freezing milk in your ice trays, you can pop out the exfoliating cubes in case of a blemish emergency. Blemishes, if squeezed, can leave a discoloration in that area of the skin. Exfoliation can help that to fade, creating a desired, monochromatic skin tone. The ice is cooling and causes vasoconstriction (shrinking of the blood vessels) and minimizes swelling or irritation from the blemish. Milk is an excellent exfoliant. As the skin ages, our natural ability to exfoliate slows down, leaving a buildup of dead skin cells. Milky Ice Cubes can help you reveal younger, brighter skin! Milk, topically, does a body good.

1 cup whole milk

Preparation and Application
Pour milk into ice cube shapes from an ice cube tray. Place into the freezer for approximately 2½ to 3 hours,

... *continued on next page*

until it's frozen and keeps its shape. Push out one milk ice cube and gently smooth all over your face in circular motions, avoiding your eye area. This should feel cooling and is a nice treat during a hot bath. Save the extra milk ice cubes in the event of a skin emergency. You can even pour 1 cup of whole milk directly into your bathwater for luxuriously soft skin.

natural beauty

Floral Healing Facial

Calendula is a beautiful flower that is closely connected with skin health. It's commonly used as a homeopathic remedy for several skin ailments, including relief of skin inflammation, bruising, slow healing wounds, and eczema. Calendula also has antifungal properties and is widely known to be antimicrobial and an excellent skin purifier. Carrots are rich in vitamin A, known for supporting acne-prone skin. Sweet almond oil is a wonderful natural moisturizer. Because calendula is antimicrobial, it's an excellent skin purifier.

3 cups water
3 carrots
4 tablespoons organic honey
2 tablespoons sweet almond oil
2 tablespoons dried calendula flowers

Preparation and Application
Cut carrots into chunks with the skin on (the nutrients are in the skins, so no need to peel, just wash well). Bring a medium pot of water to a boil. Throw in carrots. Boil

… continued on next page

until the carrots are soft, about 15 minutes. Mash them in a bowl. Add the organic honey and almond oil. Add calendula flowers. Mix until you have a smooth orange paste. Place in the refrigerator for 10 minutes to cool. It's a good idea to use a tongue depressor for application to avoid depositing the oil and germs from your fingertips onto your skin. Leave on for 10 minutes and rinse with cool water to seal the pores.

natural beauty

Creamy Décolletage

For a soft décolletage, follow this easy-to-make moisturizer to create a creamy complexion.

½ avocado
1 tablespoon honey
1 tablespoon sour cream

Preparation and Application
Mash avocado in a small bowl. Add honey and sour cream. Blend ingredients until you have a smooth consistency. Apply to your décolletage and allow the mixture to moisturize the delicate skin on your neck for 15 minutes before rinsing off with warm water. You will immediately feel a difference in the texture of your neck.

Oatmeal Exfoliation

Oatmeal has skin-soothing benefits and can be used as a gentle cleansing agent as well as an exfoliant. This beauty recipe is a wonderful and effective way to slough off dead skin cells in the gentlest way, using the naturally gritty texture of the oatmeal to massage over your décolletage. Just as important as it is to exfoliate your skin, it's important to replenish and hydrate. This beauty recipe does both!

   1 tablespoon honey
   ½ cup oatmeal, uncooked

Preparation and Application
In a small bowl, mix the honey and oatmeal. Scoop up the hydrating exfoliant and pack onto your neck and décolletage. Leave on your skin for 10 minutes and rinse with warm water.

natural beauty

Egg Protein Mask, page 66

Sugar Lips, page 96

Circulation Circles, page 137

Creamy Décolletage, page 115

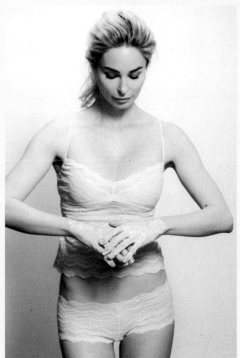

Ginger Soothing Cream,
page 145

Floral Healing Facial, page 113

Lavender Toning Mist, page 99

Brightening Mask, page 108

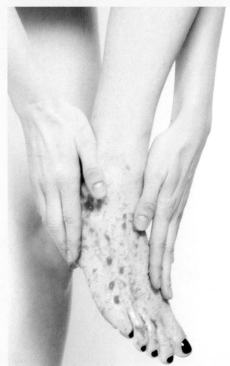

Grapefruit Feet Treat,
page 155

Cooling Mint, page 86

Beauty Byte

SHOW OFF YOUR DÉCOLLETAGE...
TASTEFULLY, OF COURSE!

- Use my Creamy Décolletage recipe (page 115) to keep your cleavage hydrated and as supple as the skin on your face.

- Remember to wear sunscreen on both your neck and chest area when exposed to sun.

- Wear a properly fitting bra with support for pretty cleavage.

- If you indulge in the more glamorous side of natural, dust some bronzer over your collarbone to highlight your neckline and dust some in the center of your chest to put your cleavage in the best light!

Healing Honey Cleavage

If your neck is slightly dry or the delicate skin of your décolletage is dull looking, this is the perfect healing beauty recipe to brighten up your cleavage! Honey is organically moisturizing and will help your skin retain its natural moisture.

1 teaspoon honey

Preparation and Application
For this simple but intensely hydrating beauty recipe, simply squeeze honey from the honey jar onto a small spoon, and then use your fingertips to apply directly to your neck and décolletage. You may want to run a bath while you leave on this hydrating décolleté mask for 15 minutes to allow the honey to moisturize your dry, dull skin. Rinse off with the warm bath water, or step into the shower and gently rub off the honey. The skin on your neck should immediately feel softer and suppler, and you notice softness to your décolletage, as well. Now, you just have to put on a V-neck, apply sunscreen, and shine!

Cucumber Neck Cream

Cucumbers soothe irritated skin. This healing vegetable improves your complexion by tightening the pores and rejuvenating your skin. It's important to give the same attention to your décolletage, and this cream is just what you need to add to your natural beauty regime. The subtle sun exposure that your décolletage suffers on a regular basis, unless you live in turtle necks, needs to be healed from any irritation from the sun's rays. It's also just a wonderful way to revitalize your neck and décolletage.

½ cucumber
1 tablespoon plain yogurt
2 teaspoons honey
1 teaspoon lemon juice

Preparation and Application
Place cucumber in a juice extractor, and pour the juice of the cucumber into a small bowl. Mix the cucumber juice with the yogurt, and then add honey and lemon juice. Mix until evenly disbursed and creamy. Apply to

… continued on next page

cleansed neck and décolletage, and allow to absorb into your delicate skin for 15 minutes. Rinse off cucumber cream thoroughly with warm water.

This is another natural beauty recipe that is best rinsed off in the shower or indulged in while bathing in the tub. However, you can just as easily gently remove with water and a soft towel.

natural beauty

In Bloom

This is an indulgent beauty recipe for a glowing neck complexion!

2 tablespoons sweet almond oil
5–6 drops rose essential oil
2 tablespoons honey

Preparation and Application
Mix oils and honey together in a small bowl until smooth and creamy. Apply to neck and décolletage. For ultimate indulgence, pour yourself a bath with 1 cup of milk and add rose petals. Leave the mask on while bathing for 15 minutes. Then cleanse with warm water. You may want to remove the mask in the shower or with a warm bath water if you have indulged with the whole milk and rose petals. However, you can also remove the mask at the sink with warm water and a soft towel. Don't remove harshly—be gentle; the skin of your décolletage is more sensitive that the skin on other parts of your body, such as your legs.

Lightening Lemon Mist

Lemon has the value of removing dead skin cells by virtue its exfoliating benefits as an astringent. The richness of vitamin C in lemons is wonderful as a skin-brightening agent. Oftentimes, we may not remember to protect our décolletage with sunscreen. As a result, hyperpigmentation (also known as sun spots or sun damage) might be present on your chest. Rosewater, also a natural astringent, tightens the capillaries underneath the skin's surface, lessening redness or inflammation. Rosewater is soothing to your skin, thereby balancing the acidity of the lemons. This beauty recipe is a great way to reveal brighter, softer skin clarity for your décolletage.

½ lemon
1 teaspoon purified water
¼ cup rosewater

Preparation and Application
In a small bowl, squeeze the juice from a halved lemon, remove the seeds with a fork and discard. Pour into a travel-size mister bottle. Simply add the purified water and rosewater to the pure lemon juice in the mister. Before showering in the morning, spray on your décolletage as

natural beauty

you would perfume. Let the mist absorb into your skin for a few minutes, and then rinse off.

For sensitive skin, I advise applying a small amount of this lightening lemon beauty recipe with a cotton ball and further diluting it with 1 tablespoon of purified water. This will lessen the intensity at which the lemon reacts with sensitive skin.

Beauty Byte

WATER: THE DRINK OF THE GODS

- Acquire the taste for water above all other beverage choices. Your skin will thank you!

- Add lemon or orange slices to your water for a citrus burst.

- Drop some raspberries into your glass to entice you to drink until you get a bite.

- Drinking water purifies your internal organs and keeps your skin clear.

- Dehydration can lead to fatigue and muscle weakness, so rely on water for energy and increased stamina at the gym or during your physical outlet of choice.

natural beauty

Cleansing Jojoba

Jojoba oil is extremely moisturizing, so it's not necessary to overindulge in this special ingredient—a little goes a long way. Rosewater cleanses the skin by removing dirt, oil, and other impurities within the skin's pores. It also tightens and clears the pores. The lactic acid in milk lightens the skin, thereby evening out skin tone. This is essential for your décolletage because it's often where sun spots are present and discoloration from sun damage is evident.

1 teaspoon jojoba oil
2 teaspoons rosewater
1 teaspoon milk powder

Preparation and Application
In a small bowl, mix ingredients together until a smooth paste is formed. Gently massage over neck and décolletage. You may even leave on the skin as a cleansing mask for 5 minutes before rinsing with warm water.

Lovely Lavender

Lavender essential oil provides aromatic relaxation and a sense of calm. Personal stress is a contributing factor on the list of toxins that cause skin issues. In addition to this import-ant benefit, lavender essential oil is an anti-inflammatory and circulatory stimulant. Lavender in the form of the powdered herb also provides these benefits. Sea clay, also known as French green clay, is the perfect base for this lovely décolle-tage natural beauty treatment. It's a true skin detoxifier and will reveal softer, revitalized skin.

2 tablespoons sea clay/French green clay
3–4 drops lavender essential oil
1 teaspoon lavender herb powder

Preparation and Application
In a small bowl, blend the sea clay with the lavender ingredients to create a creamy mixture. Apply to your neck and décolletage. Allow the mask to dry for 15 min-utes. Rinse thoroughly with warm water. As with all of the décolletage beauty recipes, this is best rinsed off in the shower stream of warm water or enjoyed in a bath and then gently removed by splashing the water on your

chest area. However, you can just as easily remove with warm water and a soft towel. Remember that the skin of your décolletage is especially delicate. This mask hardens as it dries, so be gentle when removing it.

Beauty Byte

QUICK STRESS RELIEVERS

- 20 minute power naps
- 20–30 minutes of exercise daily
- 15 minute morning meditation
- Fall asleep reading
- Adorn your home with flowers
- Indulge in a weekly bath
- Drink enough water
- Make time to connect with friends each day
- Unwind with your favorite music
- Host an at-home spa night with your favorite recipes from this book!

natural beauty

Royal Bath

What is more delightfully relaxing than a bath with milk and honey and soft bubbles topped with fresh rose petals and aromatic rose oil incense? Treat yourself like a queen!

    Petals of 3 roses, in bloom
    1 cup whole milk
    1 cup honey
    20 drops rose essential oil

Preparation and Application

Run yourself a bath of warm to hot water, depending on your preference. From the stems of three fresh roses, gently tear off the petals and place into a small bowl. Take all ingredients into your bathtub room. Play some music softly and dim the lights or turn the lights off and use candles for utmost relaxation. Pour milk and honey into your stress-reliever bath, and use your legs to stir the ingredients with the water. Lastly, add the rose petals and drops of rose essential oil.

Hot baths are wonderful for stiff muscles from stress or working out. Take deep breaths and indulge in this at-home spa experience.

Skin Detox

You skin is your body's largest organ. It has the monumental task of filtering toxins. Bentonite clay is highly effective at drawing out toxins, so this marriage of clay with your skin is a perfect match!

½ pound bentonite clay
½ pint apple cider vinegar

Preparation and Application
In a large mixing bowl, blend ingredients until a smooth paste is formed. Apply a thin coat of this healing body detox all over your legs, stomach, buttocks, arms, and décolletage. Let the heavy-duty body detox dry for 15 minutes. Rinse well in the shower. Drink a large glass of water afterward to hydrate your body and aid in the flushing out of toxins.

natural beauty

Sea Scrub

Chlorophyll, like vitamin E, is wound healing. It's also a natural deodorizer, which makes this scrub especially refreshing and cleansing for your body. In fact, you can use this scrub to exfoliate your underarms before you shave for an extra clean feeling.

    1 cup sea salt
    3–4 drops liquid chlorophyll
    3–4 drops lemon essential oil
    3–4 drops lavender essential oil
    ½ tablespoon vitamin E

Preparation and Application
Pour the sea salt into a small mixing bowl. Add the chlorophyll and essential oils to the salt and stir until you have a thick, grainy substance. Add the vitamin E for extra skin healing.

Circulation Coffee Scrub

Coffee grounds contain caffeine, which is effective for increasing circulation, and helps to reduce the appearance of cellulite when applied topically. Coffee grounds are also antibacterial. Grape seed oil is rich in antioxidants and essential fatty oils—including linoleic acid, oleic acid, and palmitoleic acid—all of which are vital to skin health. The most significant benefit of the grape seed oil is that it's readily absorbed into your skin. Ingredient absorption is key for topically applied beauty recipes. This beauty scrub penetrates your skin with antioxidants to fight free-radical damage, restore collagen and elastin, and improve circulation all at the same time.

    4 tablespoons caffeinated coffee grounds
    2 tablespoons grape seed oil

Preparation and Application
Pour caffeinated coffee grounds in a small bowl. Add grape seed oil. Mix with a tongue depressor until you have an oily, grainy scrub. Scoop up the scrub and apply to your legs, thighs, and buttocks to increase circulation and improve the appearance of cellulite. You can apply this scrub in the shower; just step out of the water stream

to apply, eliminating dead skin cells and improving circulation. Rinse thoroughly by stepping back into the stream of the shower. Go over your skin with gentle soap and shave as normal. You will get a closer shave, and your skin will feel noticeably softer from the exfoliation and absorption of the grape seed oil.

Stretch Mark Butter

This beauty recipe helps in the reduction of stretch marks, which appear on the skin as pinkish lines. The skin healing properties of vitamin E oil make this stretch mark butter superior to cocoa butter application by itself.

½ cup cocoa butter
2 tablespoons vitamin E oil

Preparation and Application
Mix ingredients in a small bowl. Apply on a regular basis to fade the appearance of stretch marks.

natural beauty

Sandalwood Scar Treatment

Acne can leave a mark, literally. Scarring from acne can be improved with this natural beauty recipe.

1 teaspoon sandalwood powder
2 teaspoons rosewater

Preparation and Application
Mix ingredients together in a small bowl until a paste forms. Apply to acne scars (or any other types of scars) and leave on for 30 minutes or longer if you have time. Gently rinse with warm water or a warm, wet washcloth. Repeat daily until you see an improvement.

Vitamin E Oil

Vitamin E oil promotes skin health on a multitude of levels. It has high antioxidant value to fight free radicals and damage done by environmental toxins and even stress. Vitamin E capsules are sold at many drugstores in the pharmacy section and can be punctured to release the oil. You can also purchase pure vitamin E oil at most health food stores in the body aisle.

2 vitamin E capsules

Preparation and Application
Pierce the vitamin E capsules to extract the pure oil and apply directly to your skin. It's particularly helpful to place on any acne scars to diminish the appearance of pock marks. Vitamin E is also wonderful in the improvement of fine lines and wrinkles, so softly rub the oil from the capsule on your forehead or around your eyes where wrinkles develop.

natural beauty

Circulation Circles

This intense body scrub is another natural beauty recipe to increase circulation. Caffeine causes vasoconstriction, which tightens the skin and lessens the appearance of cellulite. Be certain to apply this body scrub in an upward, circular motion to promote circulation and clearing of the fluid in the lymph nodes, which get clogged and create the dimpled symptoms of cellulite.

3 tablespoons olive oil
5 tablespoons coffee grounds
1 tablespoon brown sugar
3 drops orange essential oil

Preparation and Application
Mix the coffee grounds and sugar in a small bowl. Brown sugar will add a sweet smell to your body scrub. The orange essential oil will add a pleasant aroma and add to the moisturizing benefits you get from the olive oil. Add oils to your mixture and blend together until you have a grainy paste. Apply to your inner and outer thighs in an upward, circular motion to diminish the appearance of cellulite and exfoliate your skin for a better shave.

Beauty Byte

CIRCULATION BOOSTERS

→ Use a skin brush against your legs, thighs, and bottom before you shower.

→ Exfoliate your entire body with scrubs, paying extra attention to the lower half of your body.

→ Resist caffeinated drinks and carbonated beverages. They tend to cause cellulite and restrict blood flow, not to mention they create bloating.

→ Break a sweat! Exercise is a great way to get your blood flowing.

→ Niacin is a circulatory vitamin that makes you temporarily flush and increases circulation.

natural beauty

Exfoliating Body Scrub

Jojoba oil is a silky emollient that is widely known to improve stretch marks and minor wrinkles, such as crow's-feet or fine lines around the eyes. Jojoba oil has hydrating value and when mixed with sea salt becomes a luxurious exfoliant for your skin.

    4 tablespoons sea salt
    2 tablespoons jojoba oil

Preparation and Application
Mix ingredients together in a small bowl. Scoop up with your hands and apply in a massaging, circular motion to your body, including your legs, thighs, buttocks, arms, and even your stomach to scrub off dead skin cells.

Lovely Lime

Lime juice is high in vitamin C, which is a great skin brightener. If your hands are rough and dull looking, make them soft and smooth with this fun beauty recipe.

Juice from 1 lime
2 tablespoons sea salt
1 teaspoon sweet almond oil

Preparation and Application
In a small bowl, mix all ingredients. The sea salt exfoliates your hands by removing the dead skin cells, revealing softer skin. The almond oil provides much needed moisturizing, and the lime juice brightens your hands' complexion by evening out skin tone.

Anti-Aging Hand Help

Hands reveal signs of age, even before your face does. So, let's pay some attention to our hands. You will love this natural, firming antiaging beauty recipe. Protein-rich egg whites have a tightening quality and help to firm sagging or wrinkled skin when applied topically. Lemons are a natural astringent, so this beauty recipe has antimicrobial and antibacterial properties ideal for intensely cleaning our hands.

    1 egg white
    1 teaspoon lemon juice

Preparation and Application
In a small bowl, separate the egg white from the yolk by cracking the egg and pouring the yolk back and forth between the two half shells. The egg white will fall into the bowl, and you can save the yolk for my Lustrous Locks beauty recipe for radiant hair (page 64). Mix egg white with the lemon juice. Lightly saturate two cotton balls with this mixture and place over your closed eyes. Relax for 15 minutes. Remove the cotton balls. Then rinse eyes with warm water.

Shea-Olive Moisturizer

Olive oil is rich in omega-3 fatty acids, which are naturally moisturizing and softening for the skin. Olive oil also has plenty of vitamin E, which is a powerful antioxidant and is effective for soothing dry skin. Shea butter is a wonderful moisturizer and penetrates your skin deeply to restore collagen and elasticity. It's also packed with vitamins A and E, which are essential for skin health.

1 tablespoon extra-virgin olive oil
1 tablespoon shea butter

Preparation and Application
Mix both ingredients together and massage thoroughly over your hands, including the backs of your hands, which tend to be a drier area. The shea butter will melt into your hands as you apply it.

natural beauty

Rosy Avocado Oil

Avocado oil is rich in fatty acids that replenish skin with moisture. Rose essential oil not only smells divine but also has the benefits of improving circulation. It also has astringent properties. When mixed, these oils are a perfect blend for a healing hand oil treatment.

    1 teaspoon avocado oil
    ½ teaspoon rose essential oil

Preparation and Application
This is super simple! Blend ingredients together with your hands and massage all over your hands, forearms, and elbows for a healing oil treatment.

Oat Hand Scrub

Oatmeal has a gritty texture that is great for sloughing off dead skin cells. It also has moisturizing benefits! Lavender essential oil is calming and soothing and has antiseptic and antibacterial properties, making it an ideal cleansing ingredient for your hands.

> ½ cup rolled oats, plain
> 5–7 drops lavender essential oil
> 1 cup sea salt

Preparation and Application
Mix the rolled oats and sea salt in a food processor until they are blended smoothly. Scoop up and place into a small bowl, and add lavender essential oil. Scrub all over hands, forearms, and elbows for an invigorating exfoliant.

Ginger Soothing Cream

½ cup shea butter

2-inch piece of fresh ginger, juiced (equals
   approximately 1 tablespoon ginger juice)

5 teaspoons vitamin E

Preparation and Application

Place shea butter in a microwave-safe bowl and microwave
for 15–20 seconds, until melted. Add ginger juice and
vitamin E and blend with a tongue depressor. Massage
over hands, arms, and elbows while warm for a heavenly
sensation.

Avocado Hydration Hand Mask

Avocados are a wonderful source of moisture when applied topically. Dehydrated hands can be uncomfortable and even appear wrinkled much in the way that your face and neck can appear aged. Proper hydration is so important for all aspects of looking youthful and healthy. Honey is nature's moisturizer and happens to be antibacterial as well, making this a clarifying and hydrating hand mask.

  2 or 3 slices avocado
  1 tablespoon honey

Preparation and Application
Simply cut two or three slices of an avocado and blend with 1 tablespoon of honey. Gently rub the thick mixture all over the skin on your hands. Coat both the fronts and backs of your hands. Leave on for 15 minutes. Then rinse with warm water in the sink. Your hands will immediately feel smoother and softer. You may also notice a slight brightening of your hands. This is caused by the antibacterial properties of honey, which can remove dead skin cells.

Smooth as Honey

Honey is nature's moisturizer, which can rescue very dry hands.

> 1 organic egg
> 1 tablespoon honey

Preparation and Application
In a small bowl, mix ingredients. Place over hands. Dab an extra amount over your knuckles, which can appear cracked if your skin is very dry. Place a clean pair of socks over your hands and let the mixture absorb for 10 to 15 minutes. Remove socks and rinse hands thoroughly with warm water.

Sudsy Sensation

1 cup jojoba oil
½ cup honey
½ cup liquid moisturizing soap
4–5 drops lemon essential oil

Preparation and Application
In a medium bowl, mix ingredients together until a smooth, thick liquid is reached. Use a foot tub (available at beauty supply stores) or a new, unused kitty litter box as your pedicure tub. Pour in your mixture and then fill with warm water. Let your feet soak for 20 minutes. Rinse with clean water and pat dry with a towel. Enjoy your pretty feet!

natural beauty

Cocoa Buttery

Cocoa butter is amazingly moisturizing for dry feet. This combination will renew your dry or cracked toes and give you pretty feet in no time! Vitamin E oil can be found in the body aisle at most health food stores, or you can purchase vitamin E pills and puncture them with a pin to squeeze the precious oil out of it.

   4 teaspoons sweet almond oil
   4 teaspoons cocoa butter
   1 teaspoon vitamin E oil

Preparation and Application
Mix ingredients together. Smooth and massage over feet, focusing on the heels, which can take a beating.

Rosy Toes

1 cup bentonite clay
1 cup rosewater
4–5 drops rose essential oil

Preparation and Application
Mix ingredients in a medium bowl. Apply to clean feet, especially in between your toes. It's simpler and less messy to apply this beauty recipe to your feet while seated in a chair in the bathroom or sitting on the edge of an empty bathtub. Grab a book and let the rose clay foot mask dry on your feet for 20 minutes. Rinse well with warm water in the shower.

natural beauty

Softening Foot Soak

Nature's star moisturizer is honey, and when paired with whole milk, you have a recipe for beautifully soft feet.

4 cups warm water
1 cup honey
2 cups milk

Preparation and Application
In a foot tub, mix all ingredients together. Stir with your hands and let your feet soak for 15 to 20 minutes while you read your favorite book. Rinse with warm water under the bath faucet or in the shower. Dry gently with a towel.

Beauty Byte

TREAT YOUR FEET RIGHT

→ Callus removal is tricky because the skin toughens in response to being shaved down. It only grows back thicker. A natural scrub is best. Use a pumice stone to smooth out your feet and the edges of your big toe, in particular.

→ Care for your cuticles and keep them soft. Slather on oil and cream to your feet at bedtime, and cover with socks overnight. You will awaken with softened cuticles and feet.

→ File your toenails neatly. Either rounded edges or square are both lovely, as long as you commit to a shape—a recognizable shape, that is!

→ Paint your toenails with your favorite polish. Buff them if you prefer the natural look.

natural beauty

Foot Exfoliant

It's very important to regularly remove the dead skin cells from your feet. Your feet can get dull, and this recipe will help reveal softer and healthier feet. This beauty foot scrub is perfect to keep in your shower and use weekly as an exfoliant. It has a shelf life of one month, so this beauty recipe will last you for four applications. Once your skin on any part of your body is exfoliated, it's better able to absorb moisturizer. So, be ready to use the Cocoa Buttery beauty recipe (page 149) afterward!

1½ cups sea salt
¼ cup sweet almond oil
5–7 drops lavender essential oil

Preparation and Application
In a small bowl, mix coarse sea salt and sweet almond oil until a grainy paste has formed. Add lavender essential oil for a pleasing and relaxing aroma. While in the shower, scoop up the foot scrub with your hands and gently massage feet entirely, including heels and the balls of your feet, where the most pressure from walking occurs. Rinse thoroughly with water and gently dry with a towel.

Lemon Feet

Lemon is antibacterial and a natural astringent. When diluted with water, it can cleanse feet and reduce the typical itching associated with athlete's foot.

2 cups water
Juice from 1 lemon

Preparation and Application
Simply mix ingredients in a foot tub and let your feet soak for 15 minutes. Rinse with warm water and gentle soap. Pat your feet dry with a towel.

Grapefruit Feet Treat

Give you feet a warm treat during the winter. This beauty recipe has a delicious, spicy fragrance and softens the skin on your feet. Sweet almond oil helps the skin retain its own moisture; your feet need moisture, too. Brown sugar granules act as an exfoliant, removing dead skin cells. Cayenne pepper topically increases circulation, which is important for your extremities, especially in the colder months.

¼ cup brown sugar
¼ cup sweet almond oil
5–7 drops grapefruit essential oil
Pinch cayenne pepper

Preparation and Application
In a small bowl, mix the brown sugar and almond oil. Add grapefruit essential oil, and stir in cayenne pepper. In a foot tub, or a new kitty litter box as your pedicure tub, fill a few inches high with warm water. Apply scrub generously to feet, giving special attention to massaging over the tougher skin of your heels and over any calluses. Rinse thoroughly with warm water and pat dry with a towel.

8

Beauty Woes and Simple Natural Remedies

BEAUTY WOE: DULL COMPLEXION

Sun spots and other mild discolorations call for a brightening of your complexion. Beta-carotene and vitamin C are vital for maintaining a bright complexion.

Natural Remedy

Certain foods are known for improving skin discolorations internally, yet there is a special fruit that brightens your complexion internally and topically! This delicious, special fruit is the persimmon. I grew up with a huge persimmon tree in my backyard, and the antioxidant treat was sweet and given to me as dessert. My fondness for persimmons has expanded in the depths of its nutritional value for brightening one's complexion. Another helpful natural remedy is to use gentle exfoliants on a regular basis to reveal your skin's glow by sloughing off the layer of dead skin cells.

BEAUTY WOE: STRETCH MARKS

Stretch marks are a result of the skin's lack of elastin and collagen. They can also be caused by weight gain or loss.

Natural Remedy

A natural remedy for improving stretch marks is to drink lots of water. Proper hydration improves the skin's elasticity and collagen production. Cocoa butter applied topically can also help improve the appearance of stretch marks.

BEAUTY WOE:
SUN SPOTS/DISCOLORATION

Skin discoloration is also known as hyperpigmentation. Natural freckles are different from the sudden dark spots on the skin that form with age, which are known as age spots. These brown spots could be a result of sun damage that has taken years to show up. Or they could be a direct reaction to hormonal changes within the body, possibly associated with pregnancy or menstrual cycles.

Natural Remedy

Sun damage can appear on your face and on your décolletage, so it's important that your sunscreen application extends below your clavicle and covers the top of the breast area as a preventative measure. Especially for you fashion-savvy ladies who like to wear clothing with interesting necklines, make sure to always use sunscreen.

BEAUTY WOE:
BLEMISHES/ACNE

If you have an open, unhealed blemish, it can be unsightly and even painful.

Natural Remedy

Dab some toothpaste directly on your blemish. It's very drying and should calm the pimple down. You can also try some lemon juice because it's antibacterial. This is best to do at nighttime, before bed. Leave either the toothpaste or the lemon juice on for 20 minutes and then rinse off.

natural beauty

BEAUTY WOE:
VERY DRY SKIN

Hydration is key to letting your skin glow in the sun's rays, and water is the true drink of the gods. Very dry skin can be the result of the climate in your area. Regardless of the reason for your dry skin, seasonal or not, there is a simple natural remedy.

Natural Remedy

Drink plenty of water and add some raspberries or some orange slices to flavor your water and enjoy the "fruits of your labors" as you drink the last gulp. Rehydrate your skin with water, and welcome soft, smooth skin.

Another natural remedy for dry skin is to place a humidifier in your bedroom while you sleep to help your skin maintain its moisture.

Using a moisturizing soap for sensitive skin is helpful as well. Find a soap or gentle cleanser that is fragrance free and moisturizing.

BEAUTY WOE:
SCARS, ACNE OR OTHER CAUSES

Any type of scar can leave a mark that is visible on the surface of your skin. Scars can also feel tough to the touch because of the scar tissue that forms underneath the skin. Even a pimple that has been aggressively popped can create a subtle bump in the area in which it was attacked.

Natural Remedy

Vitamin E has beneficial anti-inflammatory properties. Rubbing vitamin E over a scar will help reduce the visibility of the scar over a few weeks. The act of rubbing will also decrease the tough feeling of the scar tissue under the skin, helping to break up the scar tissue.

BEAUTY WOE:
SAGGING SKIN/LOSS OF FIRMNESS

Your soft feeling can be caused by lack of muscle tone.

Natural Remedy

Improving your circulation is a key component for overall fitness. Apply either Circulation Circles beauty recipe (page 137) or Circulation Coffee Scrub (page 132) for an instant boost in your circulation. Getting your blood flowing will also prep your body for a better workout.

Leg lifts tone your inner and outer thighs; squats will help lift your buttocks. It's even fun to do at the beach with your girlfriends. Or grab your iPod and go for a run on the sand.

BEAUTY WOE:
DRY SCALP

A dry scalp can be caused by overwashing your hair, which strips your scalp of the oils that it also needs to nourish your strands of hair.

Natural Remedy

Don't overwash your hair. Allow your natural oils to nourish your scalp and make your hair shine. Once every three to five days is plenty of shampooing.

BEAUTY WOE:
DULL HAIR/LACK OF SHINE

If you hair is no longer as flexible as it once was and sort of just hangs there around your face, you may be stripping it of necessary oils produced by your scalp.

Natural Remedy

Don't wash your hair every single day, otherwise you are stripping your scalp of its natural oils, and they can't make it to the rest of your hair to provide beautiful, shiny strands. This is closely connected to the issue of a dry scalp. Condition your hair with coconut oil, and rinse easily with warm water in the shower after shampooing. In fact, you can skip the shampoo and just wet and condition it with this ultrahydrating natural recipe.

BEAUTY WOE:
CHAPPED LIPS

Chapped or cracked lips can be as painful as it looks.

Natural Remedy

Apply petroleum jelly to your lips before you go to bed at night, and you will wake up with a softer pout. Drinking water is also key to hydration, even for your lips. Be sure to use a lip balm with an SPF value in it during the day.

BEAUTY WOE:
CELLULITE

Cellulite can't be cured, but the appearance can be improved.

Natural Remedy

A circulatory stimulating body scrub is the perfect natural remedy for improving the appearance of cellulite. Drinking a lot of water is also helpful because it helps flush the toxins out of your system, clearing your lymph nodes and improving circulation. Lymph-focused massage can also help to break up the cellulite and reduce the "orange peel" look.

BEAUTY WOE:
BRUISING

Some people bruise more easily than others do, but no one wants an unsightly bruise on the surface of her skin. A lingering bruise is especially undesirable. It's a relief to know that there are things you can do naturally to expedite the healing of your bruise. If you bruise very easily, it can be the result of an iron deficiency, so it's wise to consult your doctor if this persists.

Natural Remedy

There are natural homeopathic and vitamin remedies that can heal bruises quickly. Ingest homeopathic arnica under your tongue. You can find arnica at pharmacies. Bromelain is another natural vitamin that assists with the lessening of bruising.

Conclusion

Now that you are armed with natural, pure, organic beauty recipes, your quest to reach for beauty on the natural path begins! Indulge in the luxury of your very own home spa with these creative, unique beauty recipes free of any harmful chemicals. You are a natural beauty. And you need to wield it!

Every woman who lives her life with love and kindness, physical and spiritual wellness, and generosity toward humankind is a natural beauty. There is beauty in acceptance and in nonjudgmental perspectives. It's so important to suspend judgment of those who choose an avenue different than your own to reach the same realm of beauty. Even more important, strive to embrace the virtue of finding happiness through beauty. Happiness can be found through beauty and there is, conversely, beauty in happiness. What you see in the mirror correlates to how you feel inside—your inner reflection. Be natural because that is always the best you.

Recipe Index

My Natural Beauty Recipes

My Natural Beauty Recipes

My Natural Beauty Recipes

_____
_____
_____
_____
_____
_____
_____
_____
_____
_____
_____
_____
_____
_____
_____
_____
_____
_____
_____

My Natural Beauty Recipes

_____

_____

_____

_____

_____

_____

_____

_____

_____

_____

_____

_____

_____

_____

_____

_____

_____

_____

My Natural Beauty Notes

My Natural Beauty Notes

_____
_____
_____
_____
_____
_____
_____
_____
_____
_____
_____
_____
_____
_____
_____
_____
_____
_____
_____
_____
_____

My Natural Beauty Notes

My Natural Beauty Notes